Lab Activities for the World Wide Web Annual Editions: Academic Year 2001-2002

Paula Ladd-Ruby
Arkansas State University

Ralph Ruby, Jr.
Arkansas State University

Contributing Editors

Arlieda Ries
Miami University of Ohio

R. Craig Collins
Center for Occupational
Research and Development

Scott/Jones, Inc., Publishers

PO Box 696
El Granada, CA 94018
(650) 726-2436
(650) 726-4693 Fax
scotjones2@aol.com
http://www.scottjonespub.com

**Lab Activities for the World Wide Web, Annual Editions:
Academic Year 2001-2002**
Paula Ladd-Ruby, Arkansas State University
Ralph Ruby, Jr., Arkansas State University

Copyright © 2002 by Scott/Jones, Inc.
All rights reserved. No part of this book may be reproduced or transmitted in any form without written permission of the publisher.

ISBN: 1-57676-073-1

1 2 3 Z Y X

Text Design and Composition: Paula Ladd-Ruby and Ralph Ruby, Jr.
Cover Design: Marti Sautter, Sautter Publications Services
Book Manufacturing: Von Hoffman Graphics

Scott/Jones Publishing Company
Publisher: Richard Jones
Sponsoring Editorial: Richard Jones, Patricia Myacki, and Michelle Windell
Marketing and Sales: Page Mead, Hazel Dunlap, Donna Cross, Victoria Judy
Business Operations: Michelle Robelet, Cathy Glenn, Michelle Windell

A Word about Trademarks
All product names identified in this book are trademarks or registered trademarks of their respective companies. We have used the names in an editorial fashion only, and to the benefit of the trademark owner, with no intention of infringing the trademark.

Additional Titles of Interest from Scott/Jones

Advanced Java: Internet Applications using Java
Computing with Java, 2nd Ed
Computing with Java, 2nd Alternate Ed
    by Art Gittleman

Starting Out with C++, Brief 3rd Ed
Starting Out with C++, 3rd Ed
Starting Out with C++, Alternate 2nd Ed
Starting Out with Visual Basic
    by Tony Gaddis

C by Discovery, 3rd Ed
    by L. S. and Dusty Foster

The Windows 2000 Professional Textbook
Short Prelude to Programming: Concepts and Design
Extended Prelude to Programming: Concepts and Design
    by Stewart Venit

Short Course to Active Server Pages
    by Thomas Luce

Comprehensive Windows 2000 Step-by-Step
Essential Windows 2000 Step-by-Step
    by Debbie Tice and Leslie Hardin

QuickStart to JavaScript
ShortCourse in HTML
ActiveX and the Internet
    by Forest Lin

HTML for Internet Developers
Server-Side Programming for Internet Developers
    by John Avila

The Windows 2000 Server Lab Manual 2nd Ed
    by Gerard Morris

The Complete Computer Repair Textbook, 3rd ed.
The Complete Computer Repair and A+ Reference
The A+ Self-Study Guide, 2nd Ed.
    by Cheryl Schmidt

Table of Contents

Preface VII

Lab 1 – Browser Basics 1-1

Guided Practice 1	Moving through a Web document	1-2
Guided Practice 2	Completing on-line forms	1-3
Guided Practice 3	Printing Web pages	1-4
Reinforcement 1	Working through an on-line tutorial	1-6
Reinforcement 2	Setting Bookmarks/Favorites	1-7
Reinforcement 3	Saving Web pages	1-8
Enrichment 1	Citing electronic resources	1-9
Enrichment 2	Saving and citing news stories	1-9
Enrichment 3	Following on-line stock trading	1-10
Enrichment 4	Advertising Web sites on traditional media	1-10
On-Line Reading	*A Brief History of the Internet*	1-11
Writing Practice	Comparing browsers	1-12

Lab 2 – Portal Sites 2-1

Guided Practice 1	Establishing a portal site	2-2
Guided Practice 2	Customizing the portal site	2-4
Guided Practice 3	Signing on to a chat room	2-5
Reinforcement 1	Searching for people	2-7
Reinforcement 2	Practicing establishing a portal site	2-10
Reinforcement 3	Using maps to obtain driving instructions	2-10
Enrichment 1	Going shopping on-line	2-11
Enrichment 2	Customizing a search engine	2-11
Enrichment 3	Evaluating chat rooms	2-11
Enrichment 4	Watching the weather	2-12
On-Line Reading	*Portals*	2-12
Writing Practice	Evaluating portal sites	2-13

Lab 3 – Basic Searches 3-1

Guided Practice 1	Comparing search engines	3-2
Guided Practice 2	Searching with a multi-search engine	3-4
Guided Practice 3	Developing search skills	3-6

Reinforcement 1	Reinforcing search skills	3-8
Reinforcement 2	Searching for artists	3-8
Reinforcement 3	Searching for alternative medicine	3-9
Enrichment 1	Learning about the world's fair	3-9
Enrichment 2	Locating sites for musical groups	3-10
Enrichment 3	Searching for movie stars	3-10
Enrichment 4	Purchasing an automobile	3-10
On-Line Reading	*Search Engines: How to get the best out of the Internet*	3-10
Writing Practice	Reporting on the history of the Web	3-11

Lab 4 –Electronic Mail and Mailing Lists *4-1*

Guided Practice 1	Signing up for Web-based email	4-2
Guided Practice 2	Sending file attachments through email	4-5
Guided Practice 3	Sending files through FTP	4-6
Reinforcement 1	Downloading file attachments	4-9
Reinforcement 2	Creating a distribution list for email	4-11
Reinforcement 3	Subscribing/unsubscribing to mailing lists	4-11
Enrichment 1	Practicing FTP through a Web browser	4-11
Enrichment 2	Evaluating Web-based email services	4-12
Enrichment 3	Researching and using emoticons and abbreviations	4-12
Enrichment 4	Sending Web pages through email	4-12
On-Line Reading 1	*The Core Rules of Netiquette*	4-12
On-Line Reading 2	*How Secure is Your Transaction?*	4-13
Writing Practice 1	Researching the use of email in business	4-14
Writing Practice 2	Choosing an ISP	4-14

Lab 5 –Advanced Searches *5-1*

Guided Practice 1	Using Boolean expressions and advanced search options	5-2
Guided Practice 2	Practicing Boolean expressions and advanced search options	5-3
Guided Practice 3	Using Advanced Web Search forms	5-4
Reinforcement 1	Searching specific document types and domains	5-6
Reinforcement 2	Searching in multi-search engines	5-7
Reinforcement 3	Searching for sites that identify drug abuse	5-8
Enrichment 1	Searching for downloadable resources	5-8
Enrichment 2	Searching for general trivia	5-9
Enrichment 3	Searching for graduate schools	5-10

Enrichment 4	Searching for sports trivia	5-10
On-Line Reading	*Internet Searching Strategies*	5-11
Writing Practice	Writing about distance learning courses	5-11

Lab 6 – Directory Searching *6-1*

Guided Practice 1	Category surfing for international programs	6-2
Guided Practice 2	Searching for jobs in searchable directories	6-3
Guided Practice 3	Searching in about.com	6-4
Reinforcement 1	Searching for currency conversions	6-5
Reinforcement 2	Searching for translating languages	6-6
Reinforcement 3	Searching for teen smoking and television	6-7
Enrichment 1	Searching for general and specific topics	6-7
Enrichment 2	Surfing for hobbies	6-8
Enrichment 3	Category surfing for converting measurements	6-8
Enrichment 4	Surfing for social issues	6-9
On-Line Reading	*Finding Information on the Web – Directories and Searching*	6-9
Writing Practice 1	Preparing a proposal for continuing education	6-9
Writing Practice 2	Planning a vacation	6-10

Lab 7 – Web Page Design *7-1*

Guided Practice 1	Creating a simple Web page [with HTML]	7-2
Guided Practice 2	Finding and importing clipart and animations	7-6
Guided Practice 3	Adding elements to a simple Web page	7-8
Reinforcement 1	Understanding coding and tags	7-10
Reinforcement 2	Debugging a Web page	7-10
Reinforcement 3	Creating a personal home page	7-12
Enrichment 1	Examining color psychology	7-12
Enrichment 2	Defining Web language terms	7-13
Enrichment 3	Applying the ADA to the Web	7-13
Enrichment 4	Analyzing Web sites	7-13
On-Line Reading	*The Top Ten New Mistakes of Web Design*	7-14
Writing Practice 1	Proposing a company Web site	7-15
Writing Practice 2	Creating a Web site using homestead.com	7-15

Lab 8 – Legal, Societal Issues, and Government Sites *8-1*

Guided Practice 1	Locating privacy information	8-2
Guided Practice 2	Locating government booklets	8-2
Guided Practice 3	Discovering other countries	8-4
Reinforcement 1	Identifying banned books	8-5

Reinforcement 2	Touring the Library of Congress	8-6
Reinforcement 3	Obtaining information from statistical abstracts	8-8
Enrichment 1	Politics	8-9
Enrichment 2	Exploring computer virus myths	8-9
Enrichment 3	Locating watchdog groups	8-9
Enrichment 4	Obtaining information from the Bureau of Labor Statistics	8-10
On-Line Reading 1	*What is Copyright Protection?*	8-10
On-Line Reading 2	*Web Bugs Nibbling at Computer Privacy*	8-11
Writing Practice	Writing about on-line free speech	8-11

Index **I-1**

Preface

The growth of the Internet and the quick adaptation of using the WWW as a means for accessing the Internet have brought an onslaught of new courses. The courses are being developed along a continuum from a one-hour searching course to a three-hour theory course. In each instance along the continuum there seems to be an overriding concern that there are not enough lab activities for the students to build their skills. And many instructors simply do not have time to write and test lab activities. Another concern is that the lab activities that are provided become outdated rather quickly. A third concern is that published textbooks rely too much on a specific browser. *Lab Activities for the World Wide Web* seeks to provide solutions to these concerns.

Flexibility

The purpose of this textbook is to be as flexible as possible to provide instructors and students as much Web practice as needed in the teaching and learning process. This book is not aimed at a specific academic level (freshman vs senior), but is aimed instead at the course where it is being used. *Lab Activities for the World Wide Web* can be used as a supplement in a complete Internet or WWW class where the students need to make sense of the theory that is being learned. As well, this textbook can be used as an introduction to the WWW for those classes where the focus of the course is on WWW applications.

Lab Activities for the World Wide Web is written with the purpose of adapting the text to the needs of the instructors and students. No longer is the typical student from 18 to 22 years old. Today's students span generations, from teens to senior citizens. For this reason, the lab activities in the textbook offer a wide variety of topics. As well, today's students in the same class have a variety of skill levels from beginning to advanced (and at times the students know more than the instructor). This is why some items are short and some are long, why some are simple and some are more complex.

Currency

The problem that is faced by many instructors and students is that the textbooks that are used are outdated rather quickly, sometimes before the textbook has reached the market. The World Wide Web continues to grow, which makes it both exciting and frustrating for authors, instructors and students. The answer to this challenge is to publish state-of-the-art annual editions of this text, so that there is an attempt to provide up-to-date lab activities for students to complete.

Browser Independent

Additionally, current lab texts rely either on Netscape Navigator or Internet Explorer. This textbook is written so that the emphasis is on the lab activities of the Web, not on the browser software that is used. Therefore, all of the lab activities can be completed using any browser desired. Where directions vary between the browsers, directions for both Netscape Navigator and Internet Explorer are provided.

Adaptability

This textbook can be used during lab time with the instructor present or as independent work without the presence of the instructor. While each of the lab activities involves the use of the WWW, if lab time is limited, then the students could save the Web pages for the lab activities as text files during the lab, then complete the lab activities away from the networked lab. Because each school's computer labs and availability are different, this textbook has been written to be adapted to all lab environments with ease.

A Variety of Learning Tools

To offer more flexibility in making assignments, each Lab has five types of lab activities. The *Guided Practice* lab activities provide step-by-step instructions for the students to learn new skills with screen shots as guides. The *Reinforcement* lab activities provide less guidance than the Guided Practice and are used to reinforce what was learned in the Guided Practice lab activities. However, some of the Reinforcement lab activities are written with step-by-step instructions much like the Guided Practice, when the topic requires more detailed instruction than the Guided Practice could provide. The *Enrichment* lab activities allow for the instructors or the students to select which lab activities to perform. A wide variety of subjects are covered in an attempt to appeal to as many student interests as space would allow. It is impossible to write lab activities that will be embraced by all considering the great diversity of students in today's classrooms. Each Lab contains *On-Line Reading* activities. These activities provide some theory about the topic of the Lab. If this textbook is being used as the primary textbook for an applications class, then completing the On-Line Readings as the first activity is recommended. This will give students some basic theory of what the Lab is about. Lastly, the *Writing Practice* activities provide topics for students to research and write about. Writing across the curriculum is a mainstay at educational institutions today. These activities provide a natural outlet for students to practice their writing skills while at the same time, increasing their skills for using the WWW. Each Lab has *Check It Out* boxes with Web sites that might be of interest to the students. These sites are fun and informative and have been chosen by students as their favorite sites. These sites further show the diversity of today's students in age, race, gender, nationality, and more.

A Variety of Completion Formats

Lab Activities for the World Wide Web is a bound textbook with perforated pages so the students can write answers in the text, tear out the pages, and turn them in to the instructor. As well, some of the lab activities specify that the students create a word processing document to be turned in to the instructor. Other lab activities instruct students to send the answers via email to their instructors. All of these methods of turning in the assignments are merely suggestions for the instructor who prefers to follow the textbook as it is written. However, some instructors will prefer that all answers be sent over email; others will prefer that all will be keyed in a word processor, while others prefer only the handwritten answers on the tear out pages. Again, flexibility is important with this issue. What works best for the instructor and the student is the method for turning in assignments that should be used.

Student Data Disk

To aid the students in keying their answers in a word processing document, all lab activities that provide answer space on the page, have a coordinating data file on the student data disk that is included in the textbook. The name of the student data disk is *WWW Lab Activities Data Disk*. By using the data disk the students will not have to rekey the questions, nor will the instructors have to determine what questions the students are answering. As well, the instructor can then require the students to send their answers as email attachments.

Instructor Disk

Adopters of this text will receive an instructor's disk that contains, in addition to all of the student data files, all of the text documents for the on-line readings, sample syllabi, and an answer file containing the answers to the activities that remain static.

Author Web Site

An accompanying Web site by the authors is also available for adopters of this text. The URL for that Web site is **http://www.clt.astate.edu/labactivities**. The Web site has access to the documents assigned for On-Line Reading in the case that the document has been moved or has vanished from the Web. Some lab activities require that students access the Web site to get specific instructions for completing their work. For example, many of the Writing Practice lab activities instruct the students to open the template found on the Web site where they will find specifications for writing and formatting their papers. Additional contents of the Web site will develop as instructors and students explore the many possibilities that having a presence on the Web can bring.

A Special Thanks

The following reviewers had a great effect on the ultimate form of this work. We owe a considerable debt to:

R. Craig Collins
Center for Occupational Research and Development

Don Voils
Palm Beach CC

Vickie McLaughlin
Peninsula College

Scott Craig
Lane CC

Suzanne Lybecker
Bellingham Technical College

Karen Norwood
Tarrant County CC

Arlieda Ries
Miami Univ. of Ohio

David Oscarson
Brevard CC

Terry Borst
College of the Canyons

Sam Dosumu
Colorado CC System

Anthony Klejna
SUNY Buffalo

The following professors provided feedback on previous editions of this book and we are thankful for their help:

Lyn Clark
Los Angeles Pierce College

Therese Butler
Long Beach City College

Sue Ann Wiswell
Long Beach City College

Patricia Iverson
Madison Area Technical College

Barry Kolb
Ocean County College

Rebecca Lawson
Lansing CC

Fred Hills
McLennan CC

Denise S. Leete
Pennsylvania College of Technology

Quinn Stewart
Graduate School of Library and Information Science
Univ. of Texas at Austin

Dr. Karen Norwood
McLennan CC

Michael C. Kusheba
Kilgore College

James Hearne
Western Washington Univ.

Shermane Austin
City College of NY

Jamshid Haghighi
Guilford Technical College

Robert Dick
McLennan CC

Lynda Henrie
LDS Business College

Marjean Lake
LDS Business College

Barbara Harbach
Department of Music
Univ. of Wisconsin-Stevens Point

Donna Mosier
Economics Dept.
SUNY College at Potsdam

Lab 1 – Browser Basics

Learning Objectives

Upon completion of this lab, you will be able to:
- Move through Web documents
- Complete on-line forms
- Print Web pages
- Work through a browser tutorial
- Set Bookmarks and Favorites
- Save Web pages and images
- Cite Web pages when conducting research

Topical Coverage

Upon completion of this lab, you will have explored the following topics:
- Job search
- Catalog orders
- Developmental psychologist Jean Piaget
- Web browsers
- Greeting cards
- White House tours
- Citation guides
- MSNBC news
- Stock market exchanges
- Internet history

Completion Time

The completion times for each lab activity are shown below. These are only estimated times. Learning the material is much more important than progressing quickly through the activities.

Activity	Completion Time	Activity	Completion Time
Guided Practice 1	17 min	Enrichment 1	23 min
Guided Practice 2	17 min	Enrichment 2	29 min
Guided Practice 3	17 min	Enrichment 3	Will vary
Reinforcement 1	35 min	Enrichment 4	Will vary
Reinforcement 2	23 min	On-Line Reading	35 min
Reinforcement 3	29 min	Writing Practice	Will vary

Lab 1 - Browser Basics

Guided Practice 1

This guided practice activity will help you to gain experience in moving through a Web document. You will move through the collegegrad.com page and read information about conducting job searches.

NOTE: Where directions differ between Netscape and Internet Explorer the directions for Internet Explorer will be enclosed in brackets [ ].

1. Choose **Open Page [Open]** in the **File** menu. The Open Page [Open] dialog box is shown in Figure 1-1.

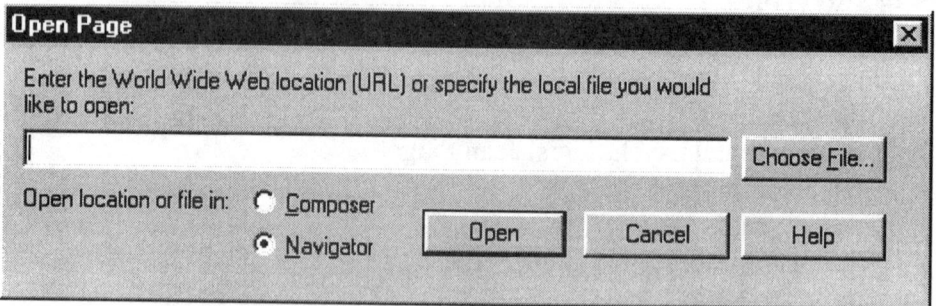

Figure 1-1 Netscape

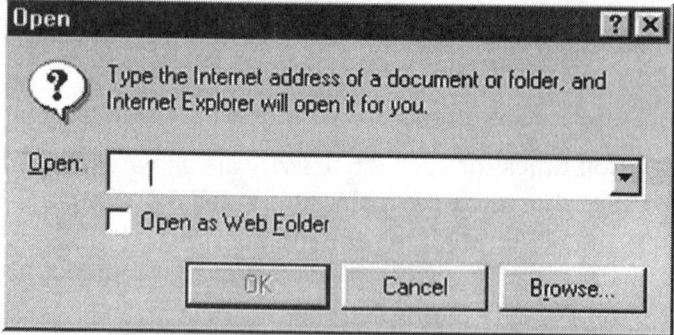

Figure 1-1 Explorer

2. Key http://www.collegegrad.com then click **Open [OK]** or press **ENTER**.

3. Click the down arrow on the vertical scroll until **E-Zine** is displayed in the left frame of the document window.

4. Click the hyperlink **E-Zine** located at the left of the screen.

5. At the E-Zine window click the down arrow on the vertical scroll bar to display additional links.

6. Click the hyperlink **Great Answers to Tough Interview Questions**

7. Read the contents of the document, then click the **Back** button on the Toolbar located in the upper left hand corner.

8. Click the **Back** button again to return to the **collegegrad.com** home page.

9. Click the **Forward** button on the Toolbar to proceed to the listing of the hyperlinks.

10. Click on three additional hyperlinks and read the contents. Record the hyperlinks and one piece of information you found helpful at each link.

Hyperlink	Helpful Information
The Truth About Resumes.	If you can't sell yourself on paper, you probably won't be able to sell yourself in person.
1.	
2.	
3.	

11. Click the **Home** button on the Toolbar to return to the start page.

Guided Practice 2

This Guided Practice will teach you to complete on-line forms by ordering a free catalog. On-line forms for requesting catalogs, placing orders, making reservations, sending greeting cards, etc. abound on the Web. Many lines in the form require information while other lines are optional. If you do not complete the required information, the form will not transmit successfully. Before giving out any private information, be sure you are at a legitimate, secured Web site.

> http://www.bloomin.com
> Send seed embedded greeting cards for a lasting memorial.

Check It Out

1. Click in the **Location [Address] Bar** as shown in Figure 1-2.

Figure 1-2 Netscape

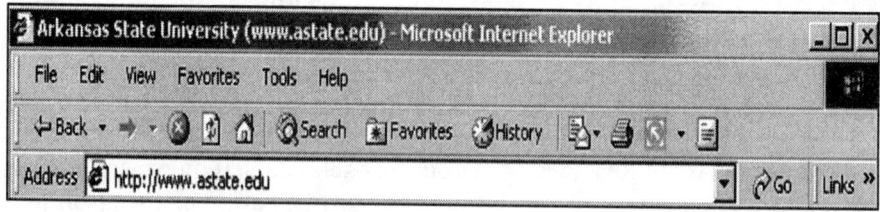

Figure 1-2 Explorer

2. Key http://www.catalogsavings.com then press **ENTER.**
3. Click a category link of your choice.
4. Click the desired catalog name.
5. Click the **Click Here!** link to order their free catalog. **Note:** If the catalog you chose has a fee, then click the **Back** button on your browser to choose another catalog.
6. As an example, the completed order form is shown in Figure 1-3 on the following page. Click or press **TAB** to move into the text boxes and key all required information.
7. When you are finished choose **Print** in the **File** menu, to print the completed order form to turn in to your instructor.
8. Click the **Order Catalog** button, or click the **Back** button on your browser if you do not to want to actually order the catalog.

Guided Practice 3

This Guided Practice will help you print Web pages. Some printers will not print pages that have a colored background with light text. This is not always a problem because you can change the color composition of the page before printing. If you are using Internet Explorer, the browser will automatically change the background color to white for printing but the display will not change. In Netscape you must change this setting manually.

1. Open the location, **http://www.piaget.org** This is the official site of the Jean Piaget Society.
2. Click on the Jean Piaget image.

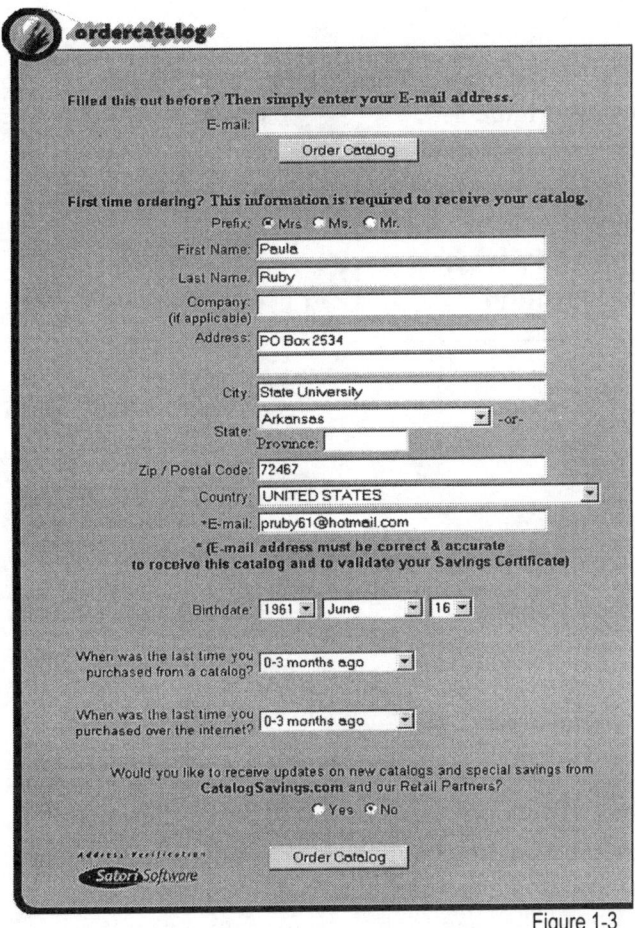

Figure 1-3

3. Click the **Print** button to print the page. If the Print button is not operational, click on the page to activate it.

1. What browser are you using?	
2. How did the page print?	

NOTE: If you are using Internet Explorer, then you are finished with this activity. The following directions are for Netscape users only.

> http://www.gamesville.com
> Play Bingo, Picturama, and more. **Check It Out**

4. Choose **Page Setup** in the **File** menu. The Page Setup dialog box is shown in Figure 1-4.

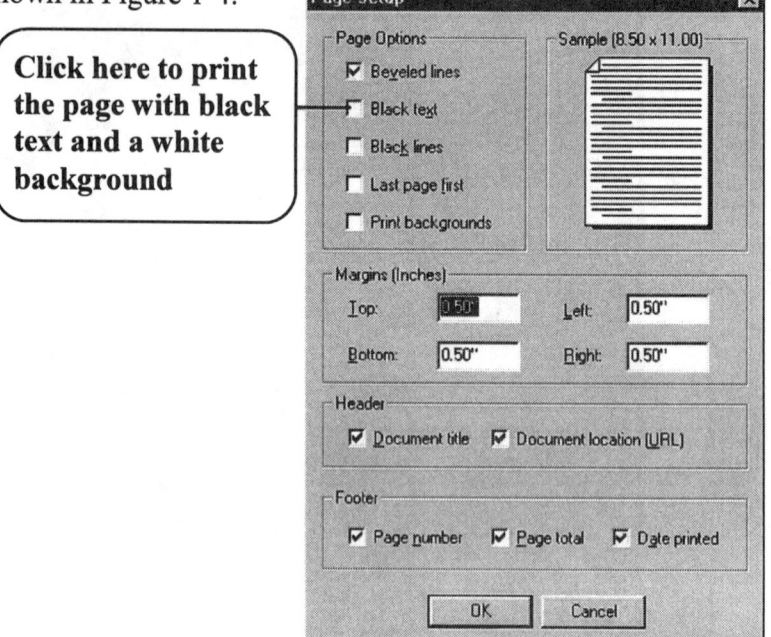

Figure 1-4 Netscape

5. The default settings are as shown. Check **Black Text** from the **Page Options**. It is important that you keep the default settings because they provide a reference to the site for future use. If any of the items checked in Figure 1-4 are not checked in the dialog box on your screen, click the box to check them, then click **OK**.

6. For **Netscape**, click the **Print** button on the Toolbar. Click **OK** to print the document. Using this setting does not change the color of the visual display, but it will print black text on a plain background. The document should now be readable.

Reinforcement 1

The WWW is a great place to learn to use the Web. Many online tutorials are available. Some are plain text and others are interactive. During this activity you are going to work your way through a browser tutorial sponsored by the National Cable Television Corporation and Tech Corps. Go to **http://www.webteacher.org** In the frame to the right is a **Web Primer** and a **Web Tutorial**. The Web Tutorial is an interactive tour through the World Wide Web. Click the **Web Tutorial** link and begin the tour. For this activity, work through the **Going Further** section. When you are finished write a memo to your instructor with examples of new things you learned about using Web browsers.

Reinforcement 2

Setting Bookmarks in Netscape or marking Favorites in Internet Explorer makes returning to often used sites quick and easy. This activity will give you practice in setting and deleting Bookmarks or Favorites and in completing on-line forms.

1. Open the location, **http://www.bluemountain.com**

2. Click the **Bookmarks QuickFile [Favorites menu item]** as shown in Figure 1-5.

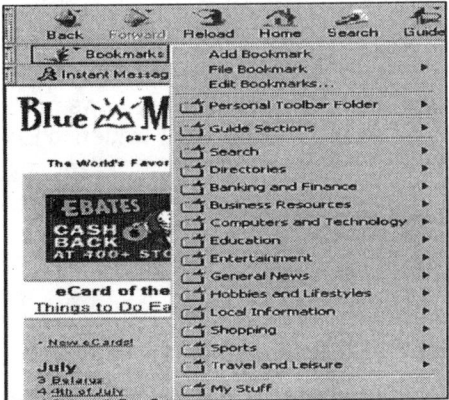

Figure 1-5 Netscape   Figure 1-5 Explorer

3. Click **Add Bookmark [Add to Favorites**, then click **OK]**.

4. Open the location, **http://www.microimg.com/postcards/**

5. Create a Bookmark or Favorite.

6. Pull down the Bookmark or Favorites list and notice the listings are now present.

7. Choose **Blue Mountain – The World's Favorite eCards – FREE!** from the listing and that page will be displayed.

8. To delete a Bookmark or Favorite, **open** the Bookmark or Favorites list. In Netscape click **Edit Bookmarks**. In Netscape and Explorer right-click on **Blue Mountain – The World's Favorite eCards – FREE!** from the list, then click **Delete Bookmark [Delete,** then click **Yes]**.

9. Return to the **Absolutely Amazing Postcards from Micro Images** bookmarked site and using the skills you learned in Guided Practice 2, send your instructor an electronic greeting.

10. On your own, add three Bookmarks or Favorites. Complete the table on the next page to indicate what three sites you marked and what the related bookmark was?

Site	Bookmark
1.	
2.	
3.	

Reinforcement 3

This activity will reinforce the skills you need to make Web pages and images permanent by saving them to disk. You can use these pages and images in other software like word processors and presentation packages.

> http://www.dawn.com
> Read daily news from
> Pakistan written in English
>
> **Check It Out**

1. Open the location, **http://www.whitehouse.gov**

2. Click on the hyperlink **Tours** located along the left hand side of the page. Scroll downward then click on the hyperlink **Online Tour of the White House**

3. Click on any one of the identified areas of the White House.

4. Choose **Save As** in the **File** menu. Insert your *WWW Lab Activities Data Disk* into the disk drive. In the **Save in** drop down list, select 3 1/2 floppy (A:) Click the **down arrow** of the **Save as type** and choose **Text File (*.txt)**. Click **Save**.

5. Right-click on the graphic of the Room, then choose **Save Image As [Save Picture As]** from the shortcut menu. Click **Save**. Minimize your browser.

6. Launch the word processor you will be using.

7. Choose **Open** in the **File** menu, then change the location to the 3 1/2 floppy (A:) drive.

8. Choose **All Files** from the **Files of type** list, then click the filename and click **Open**.

9. Delete any text that appears at the beginning of the file that was originally hyperlinks. Move to the end of the file and remove those links as well.

10. If you are using Microsoft Word, choose **Picture** in the **Insert** menu, and then choose **From File**. If you are using WordPerfect, choose **Graphics** in the **Insert** menu, then choose **From File** (if using another word processor, make the necessary adjustments).

11. Choose **All Pictures** from the **Files of type** list, change the location to the 3 1/2 floppy (A:), then click the image name. Click **Insert**.

12. Move beyond the graphic, then key your name, Reinforcement 3, and your instructor's name. Choose **Print** in the **File** menu, then click **OK**.

> http://www.phobe.com/furby/
> Conduct an autopsy on furby to see what s/he is made of.
> **Check It Out**

Enrichment 1

When you conduct research, you are required to reference your sources. You probably are already aware of citing books, journals, and newspapers, but now you must also cite electronic resources. Two common citation styles are APA (American Psychological Association) and MLA (Modern Language Association). Obtain the **citation guide template** from **http://www.clt.astate.edu/labactivities** This will provide a guide for you to use in citing electronic resources. Locate one newspaper article, one magazine article, and one Web site, then write the complete citation for each.

Resource	Citation
1. Newspaper article	
2. Magazine/Journal article	
3. Web site	

Enrichment 2

Having the ability to save a permanent copy of a Web-based document is essential because the phrase, "here today, gone tomorrow" is definitely applicable to the Web. During this Enrichment activity you will save a news story to be used later.

1. Open **http://www.msnbc.com**. Open one of the top news stories for today. Save the document to the *WWW Lab Activities Data Disk*.

2. If using Netscape, select the text of the article. Choose **Save As** from the **File** menu. At the Save As dialog box change the destination to **3 1/2 Floppy (A:)**, name the file with the extension **.txt** to make this a text file. Click **Save**.

3. Also save an associated image of the story. Right-click on the image, click **Save Image As [Save Picture As]**. At the Save As dialog box change the destination to **3 1/2 Floppy (A:)**, name the file with the extension **.gif** or **.jpeg** [Graphics Interchange Format or Joint Photographic Experts Group] to make this an image file. Click **Save**.

4. Open the document in a word processor, then insert the image (or graphic) into the file. Clean up the file so that only the text of the news story and the graphic are displayed.

5. Following the format for citing electronic sources Enrichment Activity 1, key the citation at the bottom of the document under the centered heading, Reference. Format the story and image into a presentable format and print a copy to turn in to your instructor.

Enrichment 3

One activity that has changed because of the widespread use of the Web is that of stock trading, or at least the convenience of watching the movement of stocks you own or are interested in buying or selling. Open **http://www.nyse.com** or **http://www.nasdaq.com** Select any five stocks and track the activity for three days. Obtain the **stock trade template** from **http://www.clt.astate.edu/labactivities** to record the stock activity.

Enrichment 4

The media bombards us now with Web sites related to what is being broadcast or printed from news to product advertisements. Take note for a few hours or a day until you are able to record five Web sites brought to you from broadcast or print media. Go to the sites. Did they meet your expectations based on the information you were given in the media? Will you use them often now that they have been viewed? Record your answers in the table that follows.

> http://www.sportsillustrated.cnn.com
> The ultimate in sports news.

Check It Out

URL	Expectations Met?	Use Them Again?
1.		
2.		
3.		
4.		
5.		

On-Line Reading

Go to **http:/www0.delphi.com/navnet/faq/history.html** Read through the document titled *A Brief History of the Internet*. If the document is no longer at that site, open the document from your *WWW Lab Activities Data Disk*. Locate the answers to the following questions. Compose and send an email message to your instructor with your answers, or key your answers in the correct file on your *WWW Lab Activities Data Disk*.

1. Who were three people instrumental in developing the Internet?
2. When did the Internet go online?
3. When was E-mail adapted for ARPANET and by whom?
4. When was the TCP/IP universally adopted?
5. What does the acronym, BITNET, stand for?
6. What was the first really friendly interface to the Internet and when was it developed?
7. Was Archie really named after the comic strip character?
8. Who developed the World Wide Web and when was it developed?
9. In 1993 who was the brains behind Netscape?
10. Who was the first national commercial online service?
11. What is a current trend?
12. What was the largest merger in history?

Writing Practice

To further your knowledge of working with browsers, now would be a good time to make a comparison of some of the newest versions. Open the **browser template** at **http://www.clt.astate.edu/labactivities** for the browsers to be used in the comparison and for the writing instructions for the evaluation.

> http://www.webhelp.com
> Ask questions of human Internet guides.

Check It Out

Lab 2 – Portal Sites

Learning Objectives
Upon completion of this lab, you will be able to:
- Set up a portal site
- Edit a portal site
- Change your start page
- Chat in real time
- Locate email and postal addresses
- Obtain maps and driving instructions
- Investigate on-line shopping
- Obtain the weather forecast

Topical Coverage
Upon completion of this lab, you will have explored the following topics:
- Chat events
- Hometown businesses
- Traveling
- Shopping
- Weather
- Web supersites

Completion Time
The completion times for each lab activity are shown below. These are only estimated times. Learning the material is much more important than progressing quickly through the activities.

Activity	Completion Time	Activity	Completion Time
Guided Practice 1	23 min	Enrichment 1	17 min
Guided Practice 2	29 min	Enrichment 2	23 min
Guided Practice 3	46 min	Enrichment 3	35 min
Reinforcement 1	23 min	Enrichment 4	17 min
Reinforcement 2	29 min	On-Line Reading	Will vary
Reinforcement 3	17 min	Writing Practice	Will vary

Lab 2 – Portal Sites

Guided Practice 1

The goal of many search engines and directory sites is to become your personalized gateway into the Internet. They call themselves portals. This Guided Practice activity will help you to establish a personalized portal site using NBCi.

1. Open **http://www.nbci.com**
2. Click the **My NBCi** tab as shown in Figure 2-1.

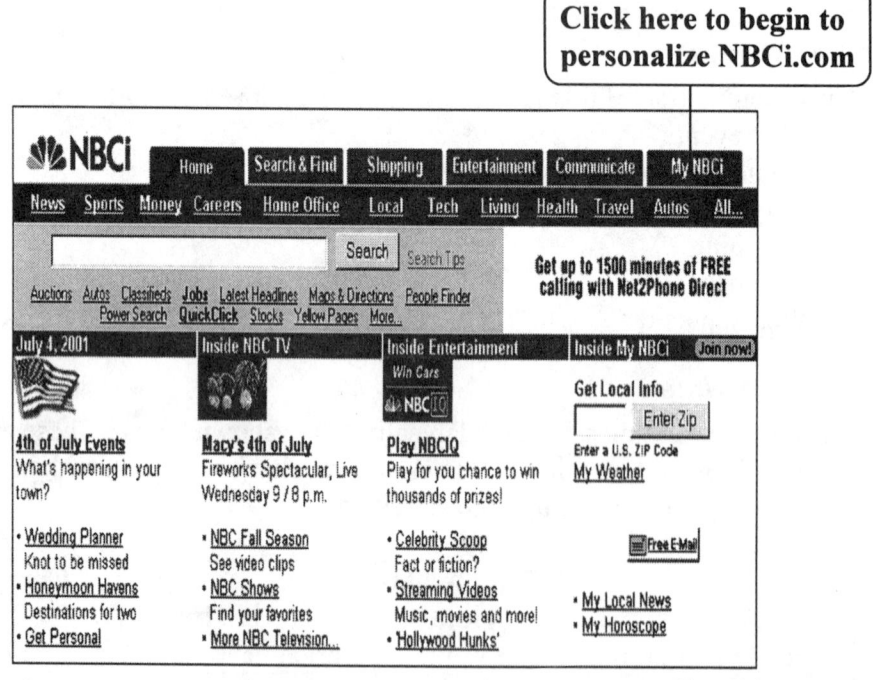

> Click here to begin to personalize NBCi.com

Figure 2-1

3. Click the **Sign Up to Personalize** link to begin.
4. Complete the sign up form by keying the required information in the text boxes or selecting from the drop down menus. Press **TAB** to proceed to the next text box. Record your member name and password somewhere safe until you commit them to memory. The completed form will be similar to the partially completed form shown in Figure 2-2.

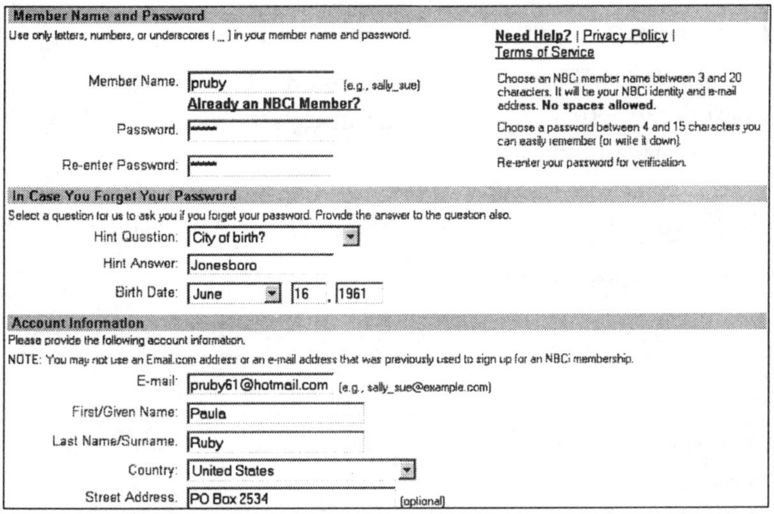

[Figure 2-2: Screenshot of NBCi member signup form showing Member Name and Password, In Case You Forget Your Password, and Account Information sections with sample data filled in (Member Name: pruby, Hint Question: City of birth?, Hint Answer: Jonesboro, Birth Date: June 16, 1961, E-mail: pruby61@hotmail.com, First/Given Name: Paula, Last Name/Surname: Ruby, Country: United States, Street Address: PO Box 2534).]

Figure 2-2

5. When the form is completed to your satisfaction, click the **Sign Up** button. It is possible the member name is already in use, so try another or select one of the suggested member names. A welcome page will be presented when a valid member name is chosen.

6. Click **Personalize:** one of the available options to further personalize such as **My Sports**. Add or remove customized content as you wish. Click **Done** when finished.

7. Click the **Personalize: Layout** link. Read the My NBCi Personalization Tips at the bottom of the screen prior to personalizing the layout. Click **Done** when finished.

8. NBCi will now be personalized. Answer the following questions.

1. What is the top news story today?	
2. What topics did you choose for the left frame?	

9. Click the **Sign Out** link.

10. Exit your browser to access your email. You will have an email message from **membership@nbci.com**

11. Read and save or print the message. Web-based email is part of a portal site. At NBCi.com it is email.com. You will sign up for Web-based email in Lab 4, so you should keep this email message to refer to later.

Guided Practice 2

If you completed Guided Practice 1, you made NBCi.com your portal site. This Guided Practice will lead you through the steps of editing the portal and making it your start page.

1. Go to **http://www.nbci.com** then click the **My NBCi** tab.
2. Click the **Sign In** link.
3. To change the My News and Info carriers click **Edit** as shown in Figure 2-3.

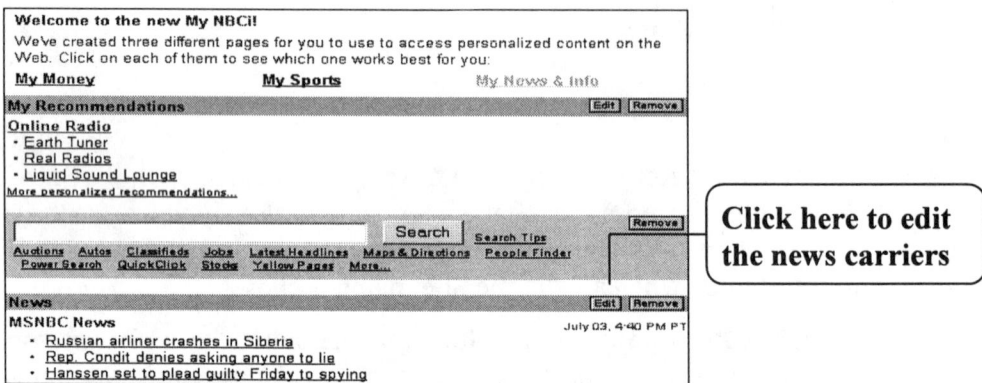

Figure 2-3

4. Click up to 25 news providers. Some have been preselected, if you do not want them, deselect. When you are finished click the **Continue** button at the end of the news list.
5. Choose the topic order then choose the number of headlines for each topic. Click **Done** when you are finished.

1. What news carriers did you choose?	
2. What topics did you choose?	

6. Continue editing **My NBCi** until you are satisfied with the contents and appearance.
7. Print the page.

2-4   Lab 2 – Portal Sites

8. Click the **Make My NBCi your start page!** Link at the bottom of the screen as shown in Figure 2-4.

Figure 2-4

NOTE: Some computer labs will not let you change the start page. Other labs will allow you to make changes, but those changes will not be saved when you log out.

9. Print the **Help Center** page that lists the directions for each browser. Follow the directions for making NBCi your start page.

10. Click the **Sign Out** link.

11. Exit your browser, then relaunch it.

12. Print your new start page.

http://www.eiu.edu/~alsiglam/
The Premier National Honor Society for Nontraditional Adult Students.

Check It Out

Guided Practice 3

One of the long-standing traditions of the Web is to be able to communicate in real-time. This is accomplished through chat. This Guided Practice activity will lead you through the steps of signing on to Yahoo! Chat. As well, you can set the Yahoo! Calendar to remind you of scheduled chats.

1. Open **http://www.yahoo.com**

2. Click **My Personalize** as shown in Figure 2.5.

Figure 2-5

3. Click the **Sign In** link at the top of the window.

4. Click the **Sign up now** link. Complete the Sign up for your Yahoo! ID page, then click the **Submit This Form** button. you may get a **Problem with Required Fields** message. If you do, make the corrections, then click **Submit This Form**.

5. To add a chat room, click the **Add Page** link.

6. At the Create a New Page window, select **Web Surfing** then click the **Create Page** button.

7. At the Web Surfing page, click **Chat** as shown in Figure 2-6.

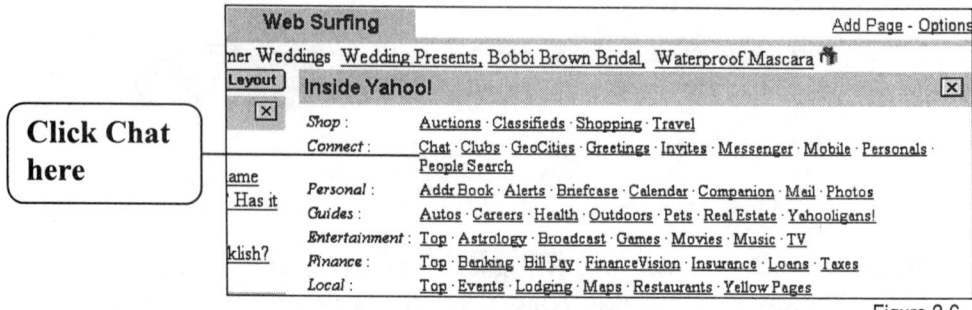

Figure 2-6

8. Click the **Help** link. What are the top 5 Questions?

1.
2.
3.
4.
5.

9. Click the **Back** button on the Toolbar to return to the main chat page.

10. Click any link in the **Featured Rooms** category. For additional choices click **Complete Room List**. Click on a chat room. You may need to load a chat applet. Have a partner pick the same chat room as you so you can practice with the ongoing discussions.

11. Read the chat discussion for a few minutes. When you are ready to join the discussion, key a message into the chat area similar to that shown in Figure 2.7 on the following page. Click **Send**.

12. Chat for as long as you are interested, then click **Exit** when you are finished.

13. Click the **Events Calendar** link located above **Featured Rooms**. This shows scheduled chats with specific topics.

14. Locate a topic of interest to you and click the **Add to my Calendar** link.

15. Add two more events on consecutive days, then print the calendar.

16. When you are finished, click the **Sign Out** link.

http://www.golfacademy.com
Improve your game with the
Swing Doctor.

Check It Out

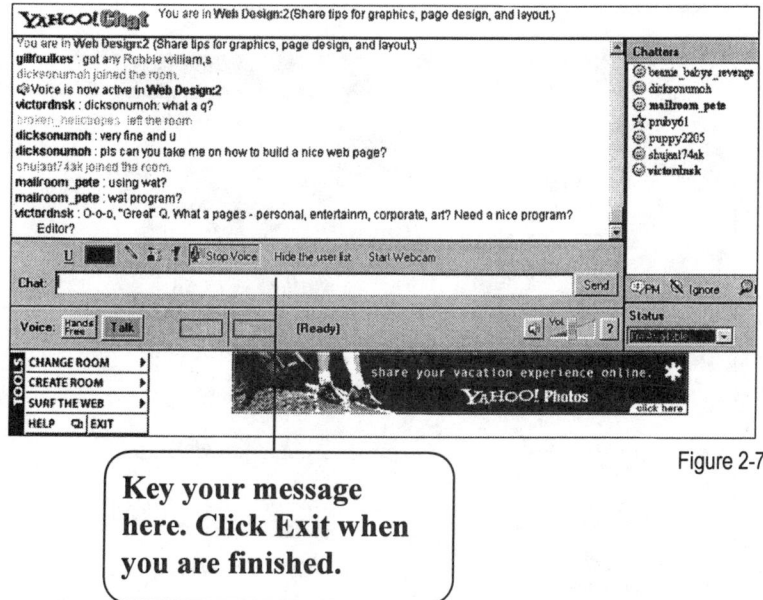

Figure 2-7

Key your message here. Click Exit when you are finished.

Reinforcement 1

Email, address, and telephone listings have become standard at portal sites. This activity will lead you through the steps for locating personal and business addresses.

1. Go to **http://www.looksmart.com** and click the **Find People** link as shown in Figure 2-8.

Figure 2-8

Click here to move to the Find People category

2. Click the **Email Directories** link, click the **General Directories** link, and then click the **Bigfoot** link. The Find People window is shown in Figure 2-9 on the following page.

Lab 2 – Portal Sites

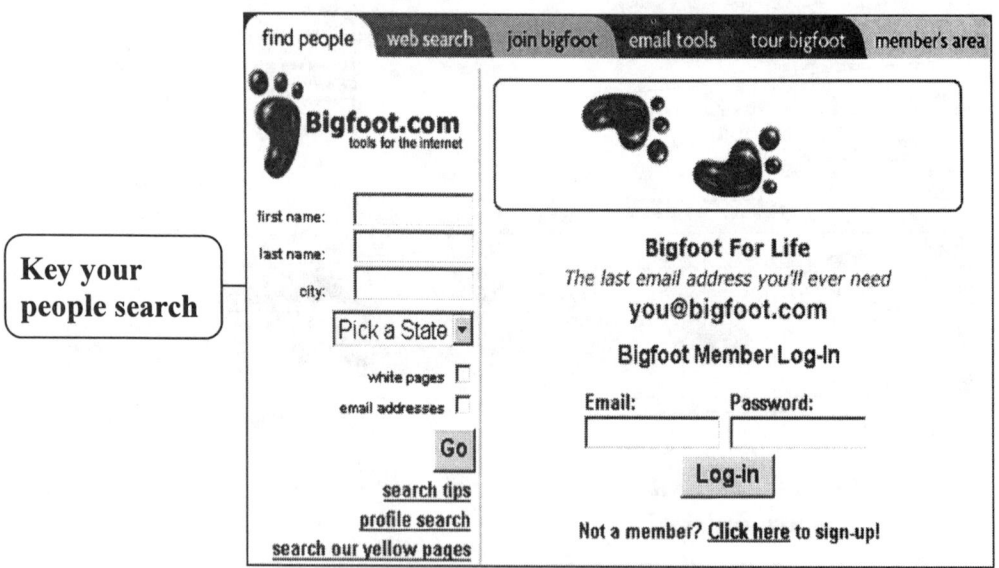

Key your people search

Figure 2-9

3. In the **find people** section, key Paula Ruby in Jonesboro, AR, then click the **email addresses** check box. Click **Go**.

1. What were the results?	
2. If no matches were found, what are the possible causes?	

4. Search for three friends or family members' email addresses.

Names	Listed in Email Database Yes/No
1.	
2.	
3.	

http://www.quicken.com/retirement/
Get your retirement savings on track.

2-8   Lab 2 – Portal Sites

5. Search for the same people in the **White Pages**.

1. Were they listed?	
2. Why or why not?	

6. Click the **search our yellow pages** link and complete the form with the following information for a category search.
7. Key Jeweler in Category.
8. Key Paragould as the city.
9. Select **Arkansas** from the select a State list.
10. Click the **Find it** button.

1. How many jewelers are listed?	

11. Return to **Search the Yellow Pages** and key restaurant in the Category and key Cracker Barrel in the Business Name.
12. Key Knoxville for the city and select **Tennessee**.
13. Click **Find it**.

1. How many Cracker Barrel Old Country Stores are displayed?	

14. Return to the **Search the Yellow Pages** form and search for three categories in or near your town, then search for three business names. Record your results.

Categories	Number of Listings	Business Names	Number of Listings
1.		1.	
2.		2.	
3.		3.	

Reinforcement 2

Practice customizing a search engine or directory for your portal site. Two suggested sites are **http://www/financialmarketsquotes.com** and **http://netscape.com** Print the home page that shows the customization when you are finished. For the **financialmarketsquotes,** enter your name or the name of a company that you wish to use. Then set clocks, set pages and set style.

Reinforcement 3

Maps and driving instructions are also part of portal sites. Follow the directions for obtaining driving instructions from your town to a distant city at the AltaVista search engine, then obtain the same driving instructions from the maps at Yahoo!

1. Open **http://www.altavista.com**
2. Click the **Directions** link from the Tools row.
3. Enter the starting and ending address, then click **Get Directions**.
4. Obtain the driving instructions, then **print**.
5. Open **http://www.yahoo.com**
6. Click the **Maps** link.
7. Enter the destination address, and then click the **Get Map** button.
8. Complete the driving instructions **To this location** link, then **print**.
9. Compare the printouts.

1. Are the directions different?	
2. Is one more detailed than the other? If yes, which?	
3. Which do you prefer? Why?	

---

http://www.historychannel.com
Find out what happened on
this date in history. **Check It Out**

Enrichment 1

On-line shopping is growing at a phenomenal rate and Yahoo! makes it easy. Go to **http://www.yahoo.com** and click the **Shopping** link under **Business and Economy**. Choose three items and follow the links to search stores on the Web. Record the three items you located, the price, and the location. Also write a statement about your impressions of on-line shopping.

Product	Price	Location
1.		
2.		
3.		
4. What are your impressions of on-line shopping?		

Enrichment 2

Based on what you learned in the Guided Practice activities go to **http://www.excite.com** and customize My Stocks, My News, My Weather, My Bookmarks, My Shopping, My Chat Events and My Horoscope. Print the start page when you are finished.

Enrichment 3

In Guided Practice 3, you went live on the Web with chat. Go to one other portal site and sign up for their chat. Answer the questions that follow.

1. What portal did you use?	
2. In what room did you chat?	
3. What are your impressions of chat?	

Enrichment 4

Go to a portal site and click the **Weather** link. Key a city in the United States and follow the prompts. Record the city, weather for today, and the URL. Next, key a location in another country and record the same information for the city.

1. City in US	
2. Weather	
3. URL	
4. Outside US	
5. Weather	
6. URL	

On-Line Reading

Go to **http:/www.zdnet.com/anchordesk/stories/story/ 0,10738,2354132,00.html** Read through the document titled *Portals*. If the document is no longer at that site, open the document from **http://www.clt.astate.edu/labactivities** Locate the answers to the following questions. Compose and send an email message to your instructor with your answers, or key your answers in the correct file on your *WWW Lab Activities Data Disk*.

About the Author: The author of this article is Ben Z. Gottesman of *PC Magazine*.

1. What does the term, Web portal, imply?
2. What do portal sites really want?
3. Who are the top two portals?
4. Who are the two lowest rated portals?
5. Why is AOL so popular?
6. What has AOL achieved?
7. Who are the runners-up for stickiness?

Writing Practice

Now that you have some experience working with portals, you should be able to evaluate them. Choose three portal sites to evaluate, then write your results. Open the **portal template** at **http://www.clt.astate.edu/ labactivities** for the items to be used in the comparison and for the writing instructions for the evaluation.

> http://www.moviefone.com
> Find out what movies are playing in your neighborhood this weekend.

Check It Out

Lab 3 - Basic Searches

Learning Objectives

Upon completion of this lab, you will be able to:
- Search using single terms
- Search using phrases
- Search using multi-search engines
- Search in different languages
- Search using case (UPPER, lower, Mixed Case)

Topical Coverage

Upon completion of this lab, you will have explored the following topics:
- Artist Salvador Dali
- Genealogy
- Breeds of dogs
- Presidential libraries
- Artists: Picasso, Van Gogh, Matisse
- Alternative medicine
- World's fair
- Musical groups
- Movie stars
- Automobiles
- Search engines
- History of the World Wide Web

Completion Time

The completion times for each lab activity are shown below. These are only estimated times. Learning the material is much more important than progressing quickly through the activities.

Activity	Completion Time	Activity	Completion Time
Guided Practice 1	23 min	Enrichment 1	46 min
Guided Practice 2	29 min	Enrichment 2	23 min
Guided Practice 3	29 min	Enrichment 3	46 min
Reinforcement 1	29 min	Enrichment 4	35 min
Reinforcement 2	46 min	On-Line Reading	23 min
Reinforcement 3	29 min	Writing Practice	Will vary

Lab 3 - Basic Searches

Guided Practice 1

This Guided Practice activity will help you to better understand how different search engines produce varied results. As well, you will determine how the search engines treat phrase searching as opposed to single-word searching. AltaVista and C|NET Search are the two search engines you will use. You will search for Salvador Dali, a Spanish painter who joined the Surrealist movement of painting in 1929.

1. Open **http://www.altavista.com**

2. To conduct a basic search, key salvador dali in the **Search for** text box, then click **Search**. The results will be similar to Figure 3-1.

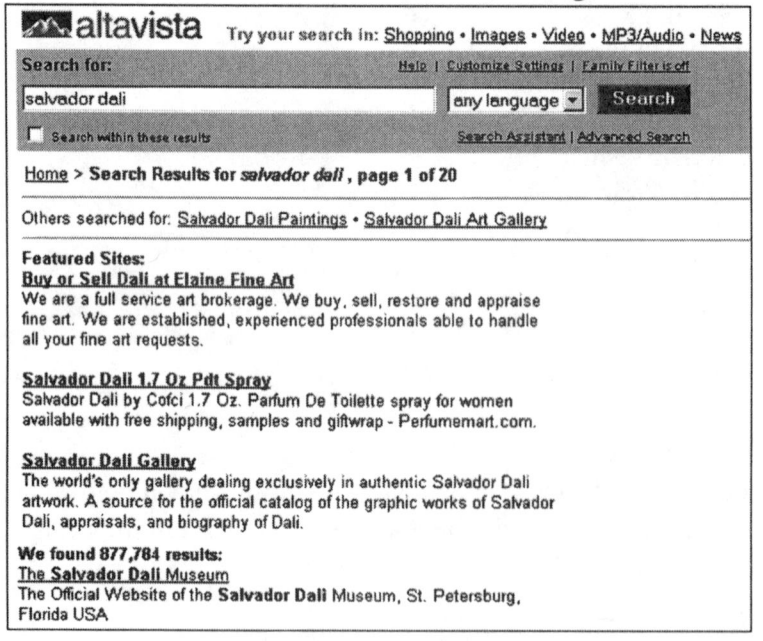

Figure 3-1

1. How many featured sites were found?	
2. How many results were found?	

3. Open **http://www.search.com/**

4. To conduct a basic search, key salvador dali in the Search text box, then click **Search**. The results will be similar to Figure 3-2 on the following page.

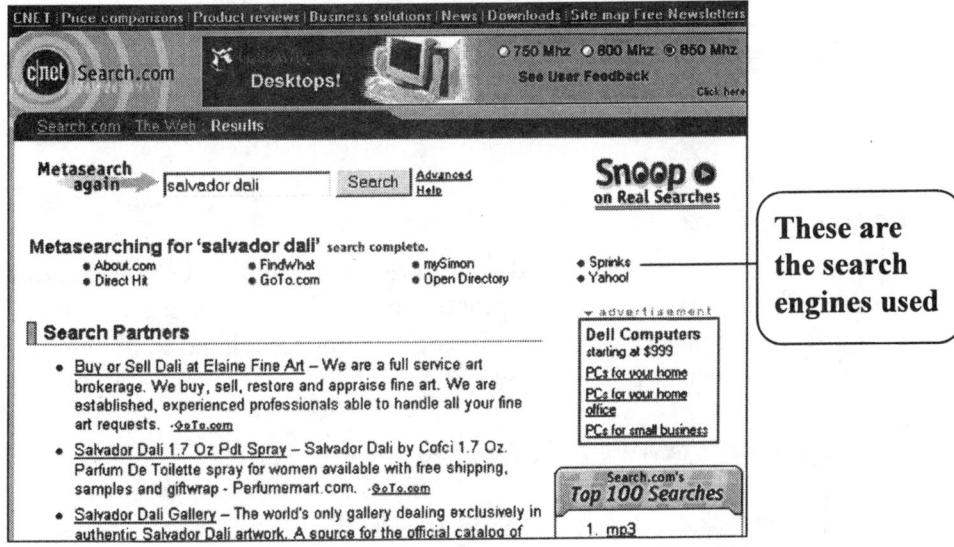

Figure 3-2

1. How many search partners were identified on the first page?	
2. During your metasearching, what search engines were used to perform the search?	

5. Open and read the documents from your search until you can find the answers to the following questions. Record the answers and the URL that you used to answer the questions.

1. What are two of Dali's most famous paintings?	
Answer	**URL**
2. What are his dates of birth and death?	
Answer	**URL**

6. Practice using both search engines for three topics of your choice. Record the three searches and one beneficial URL for each search that you were able to use in the table that follows.

Lab 3 – Basic Searches

	Search Term or Phrase	Beneficial URL
1.		
2.		
3.		

Guided Practice 2

In Guided Practice 1 you used the multi-search engine, C|NET Search. This activity will provide you with practice using ProFusion, another multi-search engine. In the process, you will locate documents related to genealogy, currently, a very popular hobby.

1. Open **http://www.profusion.com**

2. Key genealogy, then click **Find It!** The day of this search, the following results were returned as shown in Figure 3-3.

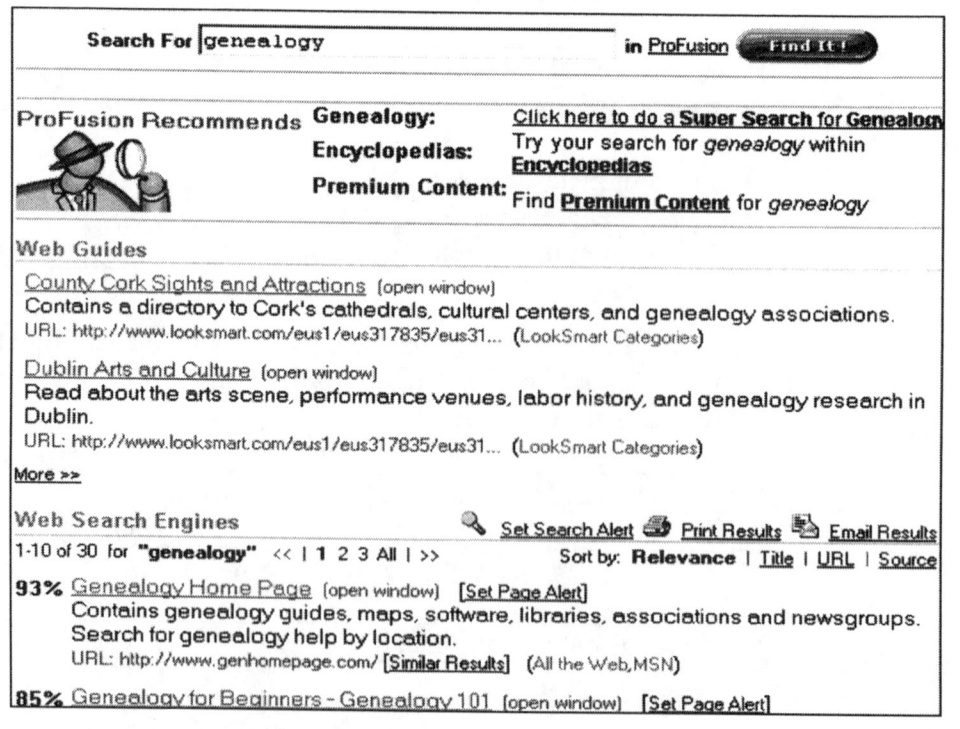

Figure 3-3

http://www.gamesville.com
Play Bingo, Picturama, and more.
Check It Out

Lab 3 – Basic Searches

3. Record the results of the first three Web Search Engines on the day of your search.

Web Search Engine	Relevance by Percent
1.	
2.	
3.	

4. Print Page 1 of the search results.
5. Click the **Back** button on the Toolbar to return to the Profusion home page.
6. Click **advanced search**.
7. Key genealogy.
8. Click the **Fastest 3** button, then click **Find It!**
9. Record the results of this search.

Web Search Engine	Relevance by Percent
1.	
2.	
3.	

10. Print Page 1 of the search results.
11. Compare the printouts. How many of the same Web Search Engines from Page 1 were returned using both searches?

1. Number of identical Web Search Engines.	

12. Click the **Back** button.
13. Perform a search using your favorite hobby as your search term. Use the same methods that you used in the genealogy search by first searching using the **Best 3** search engines, then by using the **Fastest 3** search engines.

Lab 3 – Basic Searches

14. What were the results of the first search?

1. What search term did you use?	

Web Search Engine	Relevance by Percent
1.	
2.	
3.	

15. What were the results of the second search?

1. What search term did you use?	

Web Search Engine	Relevance by Percent
1.	
2.	
3.	

Guided Practice 3

Search engines recognize search terms differently, but it is difficult to memorize the differences between the search engines. After some practice, you will develop the skills to maximize your results by using the most common searching procedures using different search engines. This Guided Practice activity will help you develop a sense of when you should key search terms in the different languages and different cases (UPPERCASE, lowercase, Mixed Case). You will be using All The Web (http://www.alltheweb.com/) and the Google search engine (http://www.google.com).

1. Open http://www.alltheweb.com

2. Perform the following searches, being careful to key the correct case. Record the number of pages and multimedia results for each search term in the table.

Search Term	Language	Pages Found	Multimedia Results
poodles	Any		
poodles	English		
poodles	French		
dalmation	Any		
dalmation	German		
dalmation	Spanish		

1. What can you conclude about the use of language and multimedia results when using All The Web?

3. Open **http://www.google.com**
4. Perform the following searches, being careful to key the correct case. Record the number of hits for each search term and case in the table.

Search Term and Case	Number of Hits
akita	
Akita	
AKITA	
scottish deerhound	
Scottish Deerhound	
SCOTTISH DEERHOUND	

Lab 3 – Basic Searches

1. What can you conclude about the use of case when using the Google search engine?

Reinforcement 1

Using the Lycos search engine, **http://www.lycos.com**, plus two more of your choice, search the phrase "presidential library" with quotation marks, then search the phrase without quotation marks. Based on the results of the search, which search engine (if any) requires phrases to be enclosed in quotation marks? Complete the table and answer the question that follows.

Search Engine	"presidential library" Number of		presidential library Number of	
	Web Sites	Articles	Web Sites	Articles
1. Lycos				
2.				
3.				

1. What can you conclude about enclosing quotation marks in the three search engines?

Reinforcement 2

Most of the time when you key a single search term you produce enough relevant documents that you can quickly locate the information you need. This activity will provide you with practice in conducting simple searches. Using the Go Network, **http://www.go.com**, conduct a simple search that will produce answers to the following questions for Picasso, Van Gogh, and Matisse. Create a word processing document to record the answers. For each artist save an image of one of his paintings and import it into the word processing document. Key the title of the painting below the image. At the end of the document, key and center the word, References, then create a complete citation for each Web document you used (refer to Lab 1 for creating citations).
1. What was the artist's date of birth?
2. What was the artist's date of death?

3. What nationality was the artist?
4. What are the titles of three famous paintings?
5. During what movement did the artist paint (for example, Surrealist)?

Reinforcement 3

Alternative medicine has made an impact on the pharmaceutical industry and the medical profession in the last half of this decade. Using two search engines, locate a helpful site from each explaining alternative medicine. Some terms that can be used are aromatherapy, acupuncture, homeopathy, and herbal healing. Answer the questions that follow for each search engine used.

Search Engine	Term(s) Used	URL of Helpful Site
1.		
This site was good because…		
2.		
This site was good because…		

http://www.blackvoices.com
Black America's Favorite Web Site. **Check It Out**

Enrichment 1

The world's fair is part of the coming together of people from all nations. A different country hosts each fair. You may have even attended a world's fair. But what exactly is the world's fair? When did it begin? What was its purpose for beginning? Using the search techniques you have learned in this lab, search for the history of the world's fair. Create a word processing document to compose your answer in paragraph format. Include an appropriate image below the paragraph. At the end of the document, key and center the word, References, then create a complete citation for each Web document you used (refer to Lab 1 for creating citations).

Enrichment 2

The Web has become a mecca for new and established musical groups and individual musical artists. Using two search engines of your choice, locate two Web sites for your favorite band or musician. Record your answers in the space provided.

Search Engine	Search Term	URLs
1.		
2.		

Enrichment 3

Conduct a simple search for three movie stars. Locate documents that will present the history of their life and the movies in which they have starred. Save the information to disk that you need to write a short composition about the three movie stars. Create a word processing document to write your composition. If you desire, include images of the movie stars in the composition. At the end of the document, key and center the word, References, then create a complete citation for each Web document you used (refer to Lab 1 for creating citations).

Enrichment 4

Conduct a simple search to prepare to purchase a new automobile. Print the document that gives the price and the features of the automobile. Next, conduct a simple search to prepare to purchase a used automobile. Print the most relevant document as well. Open the **automobile template** from **http://www.clt.astate.edu/labactivities** to record the information and to be able to make comparisons between the two.

On-Line Reading

Go to **http://www.philb.com/compare.htm** Read through the document titled, *Search Engines: How to get the best out of the Internet,* to locate the answers to the following questions. If the document is no longer at that site, open the document from **http://www.clt.astate.edu/labactivities** Compose and send an email message to your instructor with your answers.
1. What are three types of search engines?
2. How do search engines make their money?
3. How frequently is the AltaVista database updated? WebCrawler? Hotbot?

4. Is Excite a portal? Magellen?
5. What is a spam blocker?

About the Author: Phil Bradley has a background as an information professional with an Honors Degree in Librarianship. He has worked for the British Council and SilverPlatter Ltd., and is currently an independent Internet consultant. The British Library Association published his most recent book, *The Advanced Internet Searcher's Handbook*, in December 1998.

Writing Practice

Using the simple search techniques you have learned in this lab, locate three relevant documents that will enable you to write a short report on the "History of the World Wide Web." Use the search engine(s) of your choice to conduct the search. Open the **history template** from **http://www.clt.astate.edu/labactivities** for the writing instructions.

> http://www.mysterynet.com
> Solve mysteries at this fun and challenging mystery Web site.

Check It Out

Lab 4 – Electronic Mail, FTP, and Mailing Lists

Learning Objectives

Upon completion of this lab, you will be able to:
- Set up a Web-based email account
- Send and receive file attachments through email
- Use FTP with client software
- Create distribution lists for email
- Subscribe to mailing lists
- Use FTP with a browser
- Send Web pages through email

Topical Coverage

Upon completion of this lab, you will have explored the following topics:
- Email policies and conduct
- Email safety
- Etexts from Project Gutenberg
- Email services
- Net for beginners
- Emoticons and abbreviations
- Netiquette
- Security
- Digital communication
- Internet service providers

Completion Time

The completion times for each lab activity are shown below. These are only estimated times. Learning the material is much more important than progressing quickly through the activities.

Activity	Completion Time	Activity	Completion Time
Guided Practice 1	35 min	Enrichment 1	29 min
Guided Practice 2	23 min	Enrichment 2	Will vary
Guided Practice 3	58 min	Enrichment 3	Will vary
Reinforcement 1	23 min	Enrichment 4	17 min
Reinforcement 2	17 min	On-Line Readings	46 min
Reinforcement 3	17 min	Writing Practices	Will vary

Lab 4 – Electronic Mail, FTP, and Mailing Lists

Guided Practice 1

The use of Web-based email accounts has become common practice for a variety of reasons. For instance, Web-based email is free to users of the Web, and it allows you to communicate electronically from any location in the world, theoretically, instantly. All you need is access to the Web. This Guided Practice activity will help you set up a Web-based email account using the Microsoft provider, Hotmail.com.

NOTE: If you already have an account with Hotmail adjust this Guided Practice activity to set up an account through NBCi.com or Yahoo.com.

1. Open **http://www.hotmail.com** The Hotmail home page will be similar to Figure 4-1.

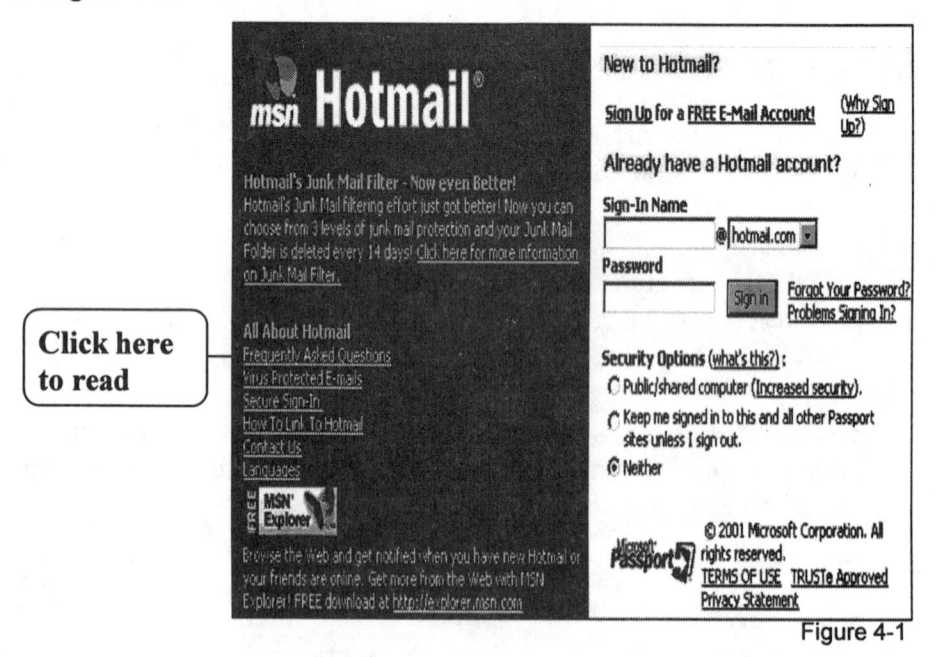

   Click here to read

   Figure 4-1

2. Click the **Frequently Asked Questions** link and read this important information. As you are reading look for answers to the following questions.

1. How much e-mail storage space is available to you?	
2. What should you do if you receive spam email?	
3. Why should you never open attachments from unknown sources?	
4. How can you protect yourself from spam mail?	
5. How do you terminate or delete your account?	

3. Click the **Home** link or click the **Back** button on your browser.

4. Click the **Sign Up** link to sign up for a free account.

5. Complete the Hotmail registration. Press **TAB** to proceed to the next text box or use the mouse to click in the text boxes. **NOTES:** Be sure to select the correct time zone. If you do not want to be listed in email directories or Internet white pages, then deselect those checkboxes.

6. When you are finished, click the **Sign Up** button. It is possible that you miskeyed or omitted a necessary piece of information or someone already has your desired sign in name. If so, correct the error as displayed, and then click the **Sign Up** button again.

7. Your Hotmail address will be displayed. Print the screen. Click the **Continue at Hotmail** button. Read the MSN HOTMAIL TERMS OF USE (TOU).

8. Click the **I Accept** button (if you do not accept the TOU, then click the **I Decline** button without completing this activity).

9. If you wish to subscribe to the WebCourier, click the boxes of topics and sites that interest you. If you choose not to subscribe, scroll to the bottom of the page. When you are finished (or do not subscribe) click the **Continue** button.

> http://www.mapquest.com
> Everything you need to plan a trip or use as a portal.
>
> Check It Out

Lab 4 – Electronic Mail, FTP, and Mailing Lists

10. If you wish to subscribe to Special Offer Newsletters, click the boxes of topics that interest you. When you are finished click the **Continue to E-mail** button. Click the **Inbox** Tab. The Hotmail Mailbox is now shown in Figure 4-2.

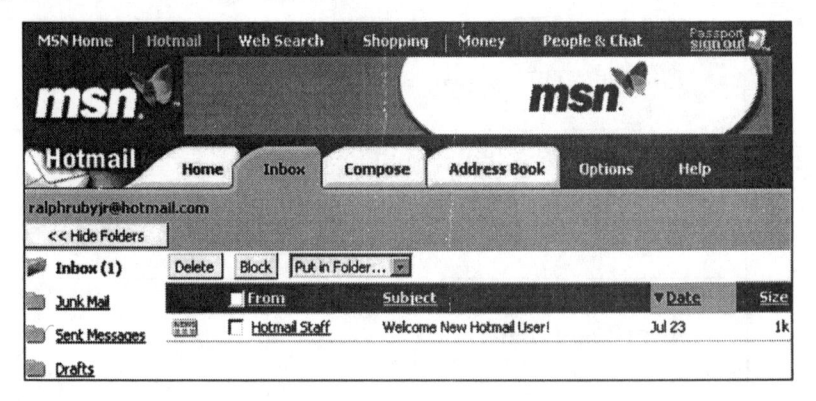

Figure 4-2

11. Notice there is mail from Hotmail Staff. Click the link.

12. Read the message, then click **Delete**.

13. Click the **Compose** Tab. Key your instructor's email address in the **To** text box. Press **TAB** and key Testing Hotmail in the **Subject** line. Click in the Message area and key your instructor a short welcome message.

14. Click the **Tools** drop-down list, and then click **Spell Check** to check the spelling of the message, then click **Send**.

15. Click **OK** at the Sent Message Confirmation window.

16. To end your email session, click **Passport** sign out in the top right OR in the bottom right portion of the window.

Guided Practice 2

This Guided Practice continues the work you began in using Hotmail.com as your Web-based email provider. You will practice sending file attachments in this activity.

1. If you are beginning a new work session, open **http://www.hotmail.com** and login with your account name and password.

2. If your instructor has replied to your welcome message, click the link and read the message. Click the **Delete** button unless you are instructed to send a reply.

> http://www.country.com
> A search engine for country music fans.
>
> Check It Out

Lab 4 – Electronic Mail, FTP, and Mailing Lists

3. Click the **Compose** Tab. Key your instructor's email address in the **To** line. Press **TAB** to move into the Subject line and key Attached document. Click the **Add/Edit Attachments** button. The Attachments page is shown in Figure 4-3.

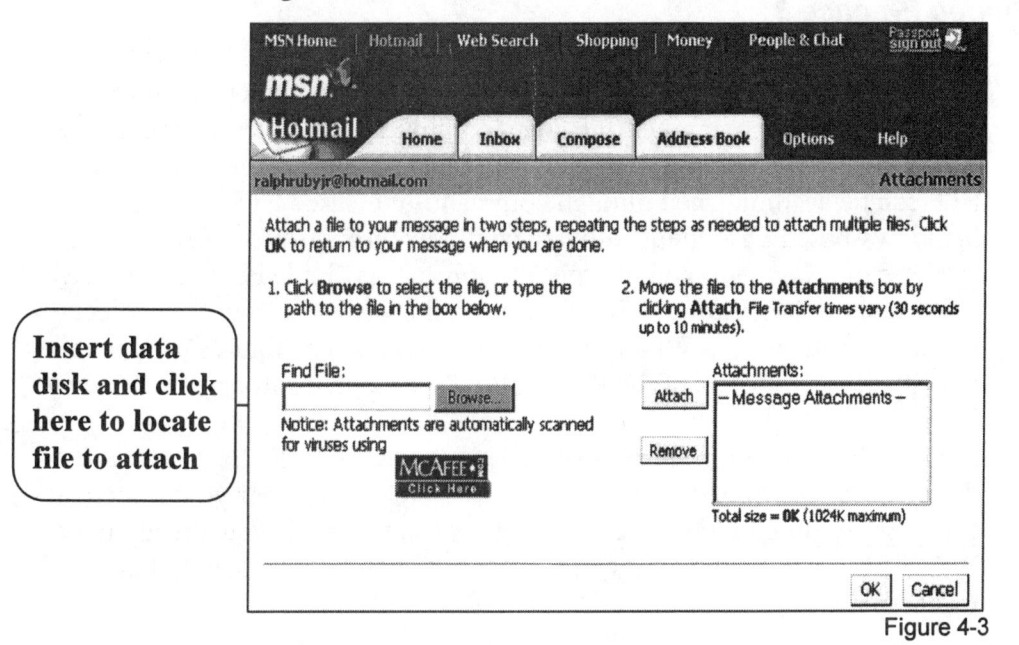

Insert data disk and click here to locate file to attach

Figure 4-3

4. Insert your *WWW Lab Activities Data Disk* into the disk drive and click the **Browse** button. The **Choose File** dialog box will be similar to the one in Figure 4-4. If necessary, change the **Look in** list to the location of your disk and choose All Files from the **Files of type** list.

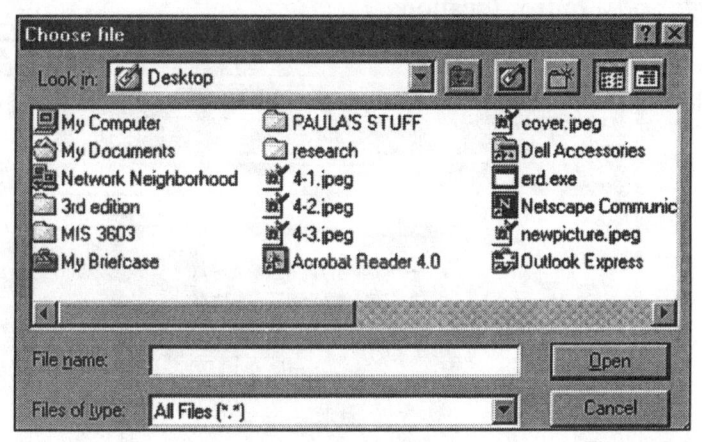

Figure 4-4

5. Click **L4 GP2.txt** then click **Open**.
6. Click the **Attach** button, and when the file name appears in the Message Attachments box, click **OK**.

7. Click in the Message area and key Here is the attached file you requested.

8. Click **Send**, and then click **OK**.

Guided Practice 3

File Transfer Protocol (FTP) is a means of transferring files between computers. When you transfer a file from a remote computer (the host computer) to your computer (the client computer) it is called downloading. When the reverse occurs, as in publishing a Web page, it is called uploading. FTP can be accomplished through command-line based FTP software, Windows-based FTP client software, and now through Web browsers such as Netscape Navigator and Internet Explorer. This Guided Practice activity will lead you through the steps of downloading an Etext from the Project Gutenberg site using a Windows-based FTP client. The figures used in this activity display the use of the WS_FTP LE 5.08 client software.

NOTE: If your computer does not already have the FTP client software, then you can download for free the appropriate software for your system at **http://www.tucows.com** The directions are as follow. If you already have WS_FTP or another FTP program on your computer, then skip to Part II of this activity.

Part I – Downloading WS_FTP (or another FTP program)

1. Open **http://www.tucows.com**

2. Click the operating system you are using from the **Tucows Downloads** box as shown in Figure 4-5. You may or may not have to choose your state and a mirror location.

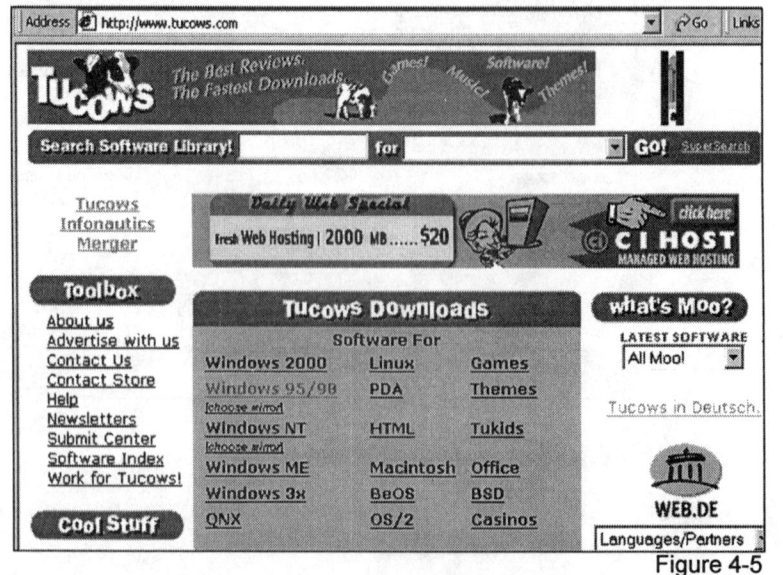

Figure 4-5

3. Click the **Download Software** link along the left frame of the w...

4. Click the **FTP and Archie** link under the heading, **NETWORK TOOLS**.

5. Scroll through the list of software and click the **Download** link for WS-FTP LE 5.08 (or another program) as shown in Figure 4-6.

```
WS-FTP LE 5.08              Freeware  ★★★★★  690.5K  Download
WS-FTP LE is a great
application that allows
remote file edits, chmods
on UNIX boxes and file
moves.
```

Figure 4-6

6. At the **File Download** dialog box, click **Run this program from its current location**, and then click **OK**.

7. If a Security Warning is displayed, click **Yes**.

8. Click **Continue**. Select **A student**, and then click **Next**.

9. Select **At school**, and then click **Next**.

10. Click **Accept**.

11. Click **OK** through the remaining prompts.

12. Click **OK** at the Congratulations Window.

Part II – Transferring a File

1. Launch the FTP client software on your computer by using the **Start** button on the Windows Task bar.

2. Complete the dialog box to include the following:

Profile Name	SunSite UNC
HostName/Address	metalab.unc.edu
USER ID	anonymous
Password	your email address

3. The completed dialog box will be similar to the one shown in Figure 4-7 on the following page.

4. Click **OK**.

5. Double-click the **[-a-]** directory on the left side of the dialog box that represents the local system (your computer; the client).

6. Click the **ChgDir** button on the right side of the dialog box that represents the host computer.

7. Key /pub/docs/books/gutenberg/ then click **OK**.

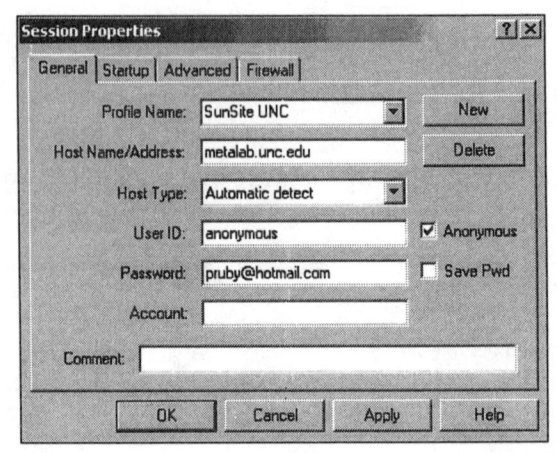

Figure 4-7

8. Click **Gutindex.99**, and then click **View**. This is the Index file for all of the documents Project Gutenberg has available as electronic texts. You will match the filename in this index to the filename in the etext99 folder. The following is an example.

The Gutenberg99 index will display the following information. From this information you obtain the Etext title of Hero Tales From American History and the filename of htfahxxx.xxx.

Mon	Year	Title and Author	[FILENAME.EXT]	####
Aug	1999	Hero Tales From American History. Lodge/Roosevelt	[htfahxxx.xxx]	1864

Using the filename.ext from the gutenberg99 index you can cross-reference to the etext99 folder to obtain the text file that you will download. The key to the cross-reference is the first five letters of the filename.

Name	Date	Size
Htfah10.txt	19990104	317917

---

9. Scroll through the list of Etexts until you locate a title you want to download. The file type should be **txt**. This example used **Hero Tales from American History**. The Filename.ext #### is **[htfahxxx.xxx]1864**. What is the title and filename of the Etext you wish to download?

1. Etext Title	
2. Filename	

10. Click the **Close** button of the **Notepad** window.

11. Double-click on the **etext99** folder and scroll through the list of Etexts until you see the filename you wrote in the previous table.

12. Click on that file, then click the left transfer arrow **[<--]** in the center of the dialog box.

13. The file transfer will begin. When the transfer is complete, the bottom line of the window will read **Transfer Complete**. Click the **Exit** button.

14. Open a word processor and open the file that was saved to your disk. A few pages of information about Project Gutenberg will be displayed. Read this information as you are scrolling through the document. When you arrive at the first page of the text, key your name above the title, and print only the first page of the text. You have now transferred a book via FTP from the Project Gutenberg site.

Reinforcement 1

> NOTE FOR INSTRUCTORS: Before the students complete this activity do one of the following:
> 1. Send them an email attachment to their new Hotmail account (one has been prepared on the instructor's data disk; filename L4 R1.txt).
> 2. Have them exchange hotmail addresses with another student and send each other an attached file.
> 3. Have them send themselves an attached file from their school account.

This activity will enhance the skills you are learning for using a Web-based email service. You are going to receive an attachment, then download it in this activity.

1. Open your Hotmail account if you are beginning a new session. If you are continuing this lab, click the **Inbox** Tab.

2. Click the link in the **From** section for the person who sent you the attachment. A message similar to the one in Figure 4-8 on the following page will be displayed.

3. Click the link of the file name (**L4 R1.txt**) as shown in Figure 4-8. A Virus Scan Result will be similar to the one in Figure 4-9 on the following page.

4. If no virus is found, click the **Download File** button. If a virus was detected, click the **Cancel** button and delete the email message immediately.

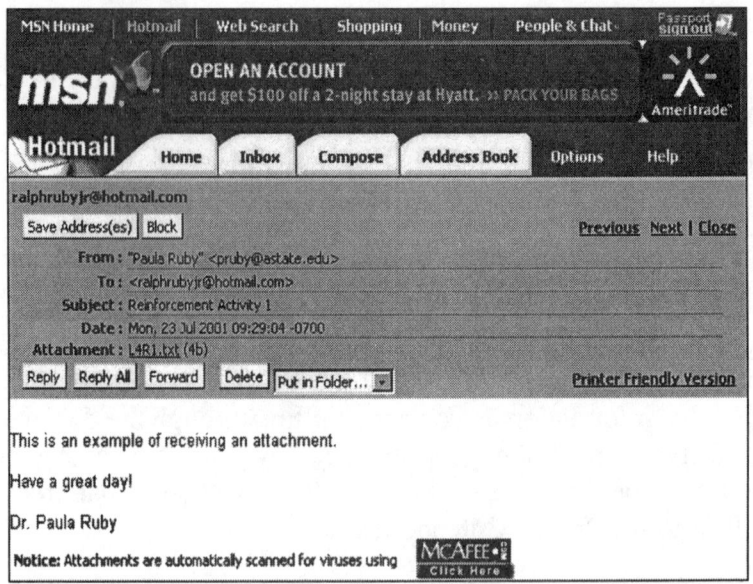

Figure 4-8

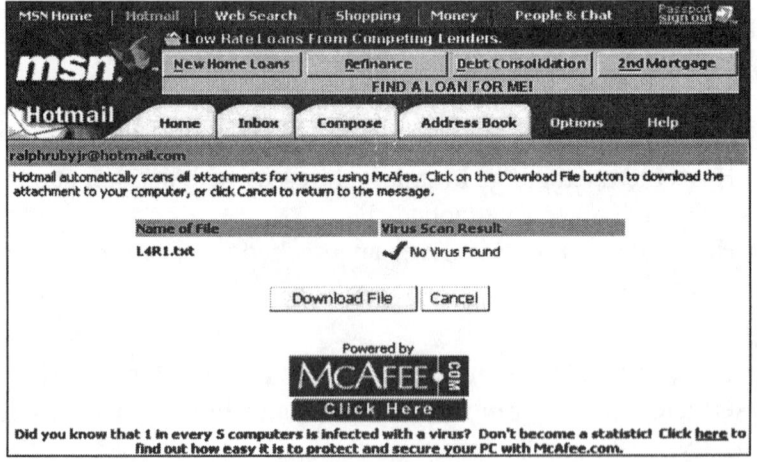

Figure 4-9

5. When the **File Download** dialog box is displayed, choose **Save this file to disk** and click **OK**.

6. When the **Save As** dialog box is displayed, choose the correct drive and click **Save**.

   NOTE: Some files will open in a new window titled, **Hotmail Attachment**. If this occurs, choose **Save As** in the **File** menu, then choose the correct name, file type, and destination, and then click **Save**.

7. If necessary, click **Close** at the **Download complete** dialog box.

8. Click the **Cancel** button, and then click the **Inbox** Tab to return to read other email.

9. In a word processor open, key your name, and print the document you downloaded to turn in to your instructor.

Reinforcement 2

Email allows you to send the same message to a group of people. Different names exist for these group emails. They may be called distribution lists, mailing groups, and group nicknames, among others. Your instructor more than likely has a distribution list for all the students in your class so when it is necessary to send the class an email message, all of the email addresses do not have to be keyed. In Hotmail, click the **Address Book** Tab, and then click the **Help** link. Follow the directions for sending a group list to five of your classmates, your family members, or a workgroup. Use the group nickname to send an email message to the group telling them they are now in your address book. Send a CC (computer copy) of this message to your instructor. Refer to Help in the mail program to send a CC.

Reinforcement 3

Mailing lists are groups of people with shared interests who communicate over email. Several types of mailing lists are available such as newsletters, announcement lists and discussion lists, moderated and unmoderated lists, digests, and more. The list you are going to subscribe to in this activity is a newsletter sponsored by about.com.

1. Open **http://www.netforbeginners.about.com**
2. Click the **Newsletter** link in the upper left corner.
3. Key your email address and other requested information then check any of the topics that you are interested in receiving. Click **Subscribe**.
4. Sign up for your special categories, then click **SUBSCRIBE**.
5. You will receive a welcome message through your email account. Print that message.
6. If you want to unsubscribe to the newsletter, follow Steps 1 and 2 above and click **Click here** below **SUBSCRIBE**.

Enrichment 1

During Guided Practice 3 you used an FTP client to obtain an electronic textbook from Project Gutenberg. With the advancements in Internet technology you can now download information that used to be available strictly through FTP directly from your Web browser. Go to **http://www.promo.net/pg/** and follow the directions for downloading an Etext through your Web browser. Save the Etext to disk and print the first page of the Etext being sure to key your name on the page you are printing.

Enrichment 2

Now that you have some experience working with Web-based email accounts, it is time to evaluate their benefits. Choose three Web-based email accounts, including Hotmail, to evaluate. Open the **email template** from **http://www.clt.astate.edu/labactivities** for the matrix to be used in the comparison. When you are finished, send the evaluation to your instructor as an email attachment.

Enrichment 3

Use a search engine to discover what emoticons or smileys and abbreviations are and how they are used in electronic communication. Compose an email message to a partner in class using primarily emoticons and abbreviations. As partners, translate each other's messages and forward or copy the translation to your instructor.

Enrichment 4

The contents of a Web document can be sent directly to anyone with an email account. Go to a site of your choice. Choose **Send Page** in the **File** menu. Key your instructor's email address, then click **Send**. Send Web pages to three friends and record the URL and your friends' email addresses.

NOTE: If you are unable to send Web pages through your email, then save the Web document as text and then send it through email as an attachment.

	URL	Friend's Address
1.		
2.		
3.		

On-Line Reading 1

Go to **http://www.albion.com/netiquette/introduction.html** Read through the document titled, *The Core Rules of Netiquette (Introduction, Rule 1 through Rule 10)*. If the document is no longer at that site, open the document from your *WWW Lab Activities Data Disk*. Locate the answers to the following questions. Compose a word processing document with your answers and send as an email attachment to your instructor.
1. What is Netiquette?
2. What is Rule 1?
    a. What analogy is made for the way humans exchange email?
    b. What should you ask before sending an email, according to Guy Kawasaki?

3. What is Rule 2?
    a. What should you do if you use shareware?
4. What is Rule 3?
    a. What should you do before you enter a new domain of cyberspace?
5. What is Rule 4?
    a. Define bandwidth.
    b. What should you ask yourself before sending copies of your messages?
6. What is Rule 5?
    a. What are you judged by?
    b. What is flame-bait?
7. What is Rule 6?
    a. Why does asking questions online work?
8. What is Rule 7?
    a. Define flaming.
    b. Define flame wars.
9. What is Rule 8?
    a. In the long run, why should you not read other's email?
10. What is Rule 9?
    a. What is the acronym, MUD?
11. What is Rule 10?
    a. How should you inform someone of a mistake?

About the Author: Virginia Shea has been a student of human nature all her life. She attended Princeton University and has worked in Silicon Valley since the mid-1980s. Ms. Shea now lives in Sunnyvale, California and has been dubbed the "network manners guru" by the San Jose Mercury News.

On-Line Reading 2

Go to **http://www.safeshopping.org/security/main.html** Read through the document titled *How Secure Is Your Transaction*. If the document is no longer at that site, open the document from **http://www.clt.astate.edu/labactivities** Locate the answers to the questions on the following page. Compose and send an email message to your instructor with your answers, or key your answers in the correct file on your *WWW Lab Activities Data Disk*.

About the Author: This site is created by the American Bar Association. It is a project of ABA Section of Business Law, Committee on Cyberspace Law, Sub Committee on Electronic Commerce, and Consumer Protection Working Group.

---

http://www.nasa.gov
A search engine for aeronautics and space research. Check out the multimedia gallery.

Check It Out

1. What is SSL?
2. What tells you a secure site is entered from the Web address?
3. What is SET?
4. What is "Theft of identity?"
5. What is an easy way to create a memorable password?
6. How should you record your password?
7. To place an online order what are the only two pieces of information that are required?
8. What should you NOT do if you receive unsolicited email?
9. How can you protect your computer from viruses?

Writing Practice 1

Bill Gates published a book titled *Business @ the Speed of Thought*, and in it outlined 12 steps for making digital information flow an intrinsic part of a company. Step 1 insists that communication flow through email. Use the WWW to gather information both in support of and against this step. Obtain the **Gates template** and specific instruction sheet from **http://www.clt.astate.edu/labactivities** to complete this activity.

Writing Practice 2

As the growth of the home PC market continues, more and more of us will be connecting to the Internet at home. The first step in doing this is to locate a reliable Internet service provider (ISP). This writing practice activity will help you to evaluate the ISPs in your area. As well, you will gain experience in writing operating instructions for other computer users. Open the **ISP template** at **http://www.clt.astate.edu/labactivities** to obtain the writing instructions for this activity.

---

http://www.stonepages.com/England/englandmain.html
View the Stones of England site.
Check out the Avebury site.

Check It Out

Lab 5 - Advanced Searches

Learning Objectives

Upon completion of this lab, you will be able to:
- Search using Boolean expressions
- Search using Advanced Search forms
- Search using symbols
- Conduct title and domain searches

Topical Coverage

Upon completion of this lab, you will have explored the following topics:
- United States Treasury
- President John F. Kennedy
- President George W. Bush
- Ethics
- Famous quotations
- Drug abuse
- Web resources
- Business trivia
- University requirements
- Sports trivia
- Search strategies
- Distance learning

Completion Time

The completion times for each lab activity are shown below. These are only estimated times. Learning the material is much more important than progressing quickly through the activities.

Activity	Completion Time	Activity	Completion Time
Guided Practice 1	23 min	Enrichment 1	29 min
Guided Practice 2	23 min	Enrichment 2	35 min
Guided Practice 3	23 min	Enrichment 3	46 min
Reinforcement 1	23 min	Enrichment 4	46 min
Reinforcement 2	29 min	On-Line Reading	Will vary
Reinforcement 3	35 min	Writing Practice	Will vary

Lab 5 – Advanced Searches

Guided Practice 1

This Guided Practice activity will help you conduct a search using the Boolean expression AND, as well as limiting the search to a range of dates for the AltaVista search engine. The specific question you will seek an answer to is "What is the history of the United States Treasury department and bank notes?"

1. Open AltaVista at **http://www.altavista.com**
2. Click the **Advanced Search** link and key the Boolean expression, "united states treasury" AND "bank notes" as shown in Figure 5-1.

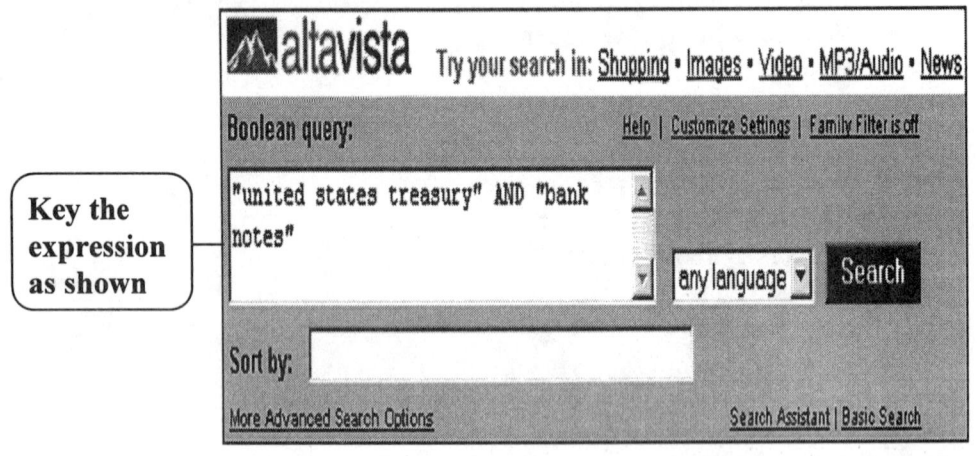

Key the expression as shown

Figure 5-1

3. Click **Search**. Your search found?

Number of pages.	Number of featured sites.	Number of results.

4. Limit your search **by date range:** of 01/01/00 to today's date, then click **Search**. How many documents were found?

Number of pages.	Number of featured sites.	Number of results.

5. Practice searching with AltaVista using Boolean expressions with and without range of dates for three additional search terms. Record all of your results in the table below.

Search Term	Results Found without Range of Dates	Results Found with Range of Dates
1.		
2.		
3.		

Guided Practice 2

In Guided Practice 1 you used the Boolean term AND. This Guided Practice activity will give you practice using symbols. AND is represented by the **&** symbol, OR is represented by the **|** symbol, and NOT is represented by the **!** symbol. You will explore the conspiracy theories surrounding the death of President John F. Kennedy. Record the results of the searches in the table that follows the step-by-step directions.

1. Open AltaVista at **http://www.altavista.com**

2. Click the **Advanced Search** link in the search area.

3. In the Enter boolean expression: text box, key (John F. Kennedy | JFK) & assassination | conspiracy theory, then click **Search**. Record the results of your search in the table on the following page.

4. Scroll back to the top of the page to the Boolean Query box and refine the search by removing the | in front of conspiracy theory and replacing it with &! (you will be removing OR and replacing it with AND NOT). Click **Search** and notice the considerable drop in hits. Record the results in the table on the following page.

5. Refine the search again, by removing the ! (NOT) so the & (AND) remains. Notice the results were drastically lower. Record the results in the table on the following page.

> http://www.harmony-central.com
> The Internet Resource for Musicians! **Check It Out**

Lab 5 – Advanced Searches   5-3

Search Terms	Results
1. (John F. Kennedy | JFK) & assassination | conspiracy theory	
2. (John F. Kennedy | JFK) & assassination &! conspiracy theory	
3. (John F. Kennedy | JFK) & assassination & conspiracy theory	

6. Construct three more searches with different topics following the Kennedy outline using the Boolean symbols. Record all of your results (hits) in the table below.

Search Terms	Results
1.	
2.	
3.	

Guided Practice 3

With the popular search engines today, you can use the Advanced Search forms to make your searching easier. This activity will guide you through the Excite search engine using the Advanced Search link.

1. Open **http://www.excite.com**
2. Scroll to the end of the page and click the **Advanced Search** link as shown in Figure 5-2.

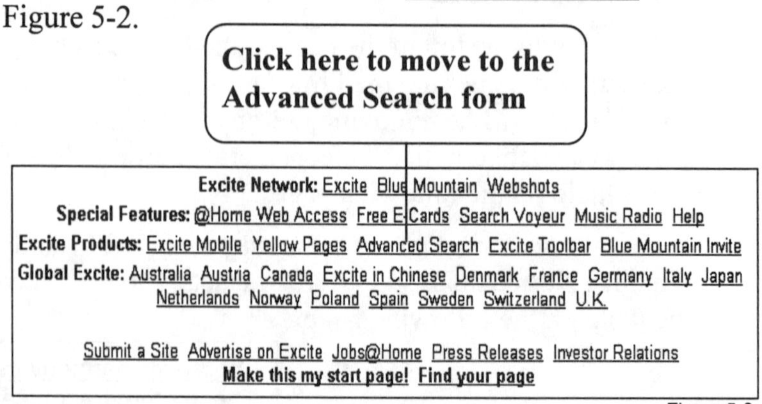

Click here to move to the Advanced Search form

Excite Network: Excite Blue Mountain Webshots
Special Features: @Home Web Access Free E-Cards Search Voyeur Music Radio Help
Excite Products: Excite Mobile Yellow Pages Advanced Search Excite Toolbar Blue Mountain Invite
Global Excite: Australia Austria Canada Excite in Chinese Denmark France Germany Italy Japan Netherlands Norway Poland Spain Sweden Switzerland U.K.

Submit a Site Advertise on Excite Jobs@Home Press Releases Investor Relations
Make this my start page! Find your page

Figure 5-2

3. To construct the search, in Step 1 choose **Must Have** from the first row of buttons. Key george w bush in the text box.

4. In the second row, the Results box **Must Have** the word president.

5. In the third row, the Results box **Must Not Have** the name al gore. The completed form will be as shown in Figure 5-3.

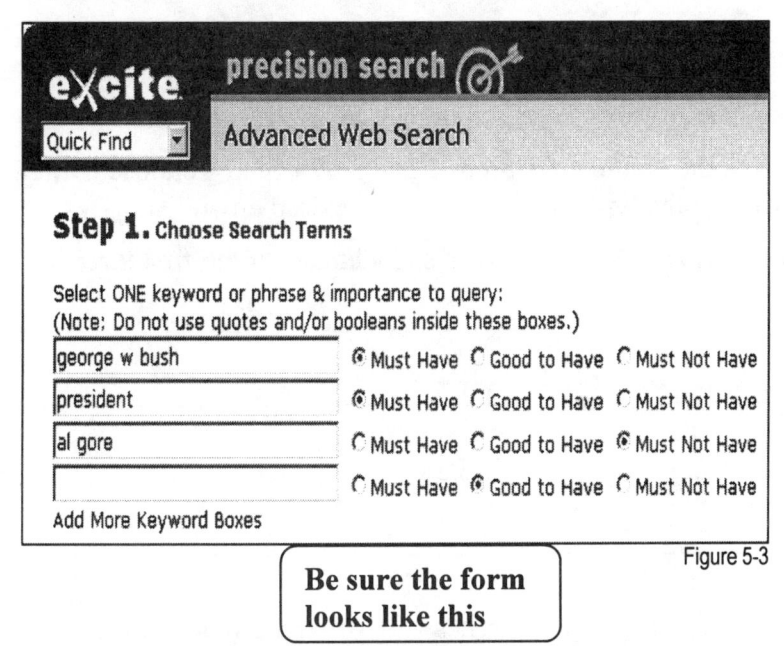

![Figure 5-3: Excite precision search Advanced Web Search form showing Step 1 with "george w bush" set to Must Have, "president" set to Must Have, and "al gore" set to Must Not Have.]

Be sure the form looks like this

Figure 5-3

6. Scroll to the end of the form and click **Search**. Write the document titles for the first three hits.

1.	
2.	
3.	

7. Click the **Modify Your Advanced Web Search** link.

8. Accept the defaults for Steps 2 and 3. For Step 4 choose **United States (.gov)** for the type of domain from the drop down list.

9. Click **Search**. Write the Web site address for the first three hits.

1.	
2.	
3.	

10. Click the **Modify Your Advanced Web Search** link.

11. Accept the defaults for Steps 2 and 3. For Step 4 choose **United States (.org)** for the type of domain from the drop down list.

12. Click **Search**. Write the Web site address for the first three hits.

1.	
2.	
3.	

13. Construct three more searches with different topics following the search you completed using the Advanced Search Form. Print the first page for each of the searches.

Reinforcement 1

Many search engines have the capability to search for specific parts of documents or specific domains. You can do this type of searching in the Search text box without completing an Advanced Search form. This Reinforcement activity will give you practice in conducting title and domain searches using the NBCi search engine. You will use the title, ethics, and edu, gov, and org domains. Record your results in the table that follows.

1. Open **http://www.nbci.com/**

2. Key title:ethics in the Search text box. Click **Search**.

3. Move the insertion point to the end of the Search text box and press the **SPACEBAR** then key +domain:edu Click **Search**.

4. Return to the Search text box, delete the edu and key gov to search only for government documents. Click **Search**.

5. Return to the Search text box, delete the gov and key org to search only for the non-profit organization domain.

1.	Write the Web site address for the first three hits for **title only**.
2.	Write the Web site address for the first three hits for **title and edu**.
3.	Write the Web site address for the first three hits for **title and gov**.
4.	Write the Web site address for the first three hits for **title and org**.

6. Practice searching with NBCi using title and domain searches for three additional search terms. Record what search terms you used.

Search Terms	First Web Site Address
1.	
2.	
3.	

Reinforcement 2

Using a multi-search engine such as ProFusion.com, Search.com, or Ask.com conduct searches to find who is given credit for coining the following quotations. Record your results in the table following the quotations.

1. Dream as if you'll live forever. Live as if you'll die today.
2. The important thing is not to stop questioning.
3. You must do the thing you think you cannot do.
4. Life can only be understood backwards, but must be lived forwards.
5. If you can dream it, you can do it.

Quote	Author	Search Terms Used
1.		
2.		
3.		
4.		
5.		

Reinforcement 3

The Web is a great resource for locating information on various drugs, both legal and illegal. It is also a quick way to research support groups when you discover that someone you know is using an illegal drug. Assume that someone you know is taking crystal meth, and you are not familiar with that drug. Using the skills you have learned thus far, first determine what crystal meth is (is it a slang word for another drug?), then determine what to do when you discover that someone you know is taking it. Create a word processing document to compose your answer in paragraph format. At the end of the document, key and center the word, References, then create a complete citation for each Web document you used (refer to Lab 1 for creating citations).

Enrichment 1

Many Web resources are free for the asking on the WWW. Using the advanced searching techniques you have learned thus far, locate one site for each of the "downloadable" resources on the following page.

http://www.ricksteves.com
Travel Europe through the Back Door

Check It Out

1. Free clipart images from org sites.	
2. Free videos from edu sites.	
3. Free video player.	
4. Free sounds.	
5. Free music.	
6. Free sheet music for an instrument (piano, violin, saxaphone, etc.).	

Conduct a search to find the answer to the following questions. Write a memo to your instructor with your findings. Make a reference list to attach to the memo (see Lab 1). Are free sites ethical? How do the artists get paid? What is MP3? The last paragraph(s) should be your opinions on this topic.

Enrichment 2

The Web is a good place for trivia lovers. You can find the answers to almost any question using the search techniques you have learned thus far. Using search engines of your choice, locate answers to following questions.

1. What connection did Hewlett-Packard have with Disney in the early days of the business?	
2. What famous baseball player was an executive of the company, Chock Full O'Nuts?	
3. What is the oldest Chicago business still in operation?	
4. What is Netscape's code name?	

Enrichment 3

Searching for universities and academic programs has greatly enhanced the process of locating a school to attend. Assume that you will be looking for a baccalaureate program or a graduate program for the next school term. Using the search techniques you have learned, locate three university degree programs you would be interested in finding more information. Go to the university sites and find the admittance requirements, fee scales, and on-line application forms. Mark the pages as Bookmarks or Favorites so you can easily refer to them and make an educated decision about which school to attend. Create a word processing document outlining the pros and cons of each school to help you make an informed decision.

Enrichment 4

The World Wide Web is a mecca for sports news. Using the search skills you have developed answer the following sports trivia questions.

1. Who holds baseball's major league record for most consecutive games played?	
2. Who is the leading scorer of the WNBA and what is her average per game score?	
3. Who is the most recent five-time winner of the Tour de France?	
4. What car did Nascar driver, Dale Earnhardt drive?	
5. During what years was the All American Girls Professional Baseball League operational?	
6. In 1931 what woman struck out Babe Ruth and Lou Gehrig?	

On-Line Reading

Go to **http:/www.rice.edu/Fondren/Netguides/strategies.html** Read through the document titled *Internet Searching Strategies*. If the document is no longer at that site, open the document from **http://www.clt.astate.edu/labactivities** Locate the answers to the following questions. Compose and send an email message to your instructor with your answers, or key your answers in the correct file on your *WWW Lab Activities Data Disk*.

1. What are the four steps to formulating a strategy?
2. What site is good for searching many Internet resources?
3. What site is good to search several WWW indexes at the same time?
4. What site is good to find an email address?
5. What does the default OR do with a phrase?
6. What does the AND or + do in a search?
7. How do you search for a phrase?
8. Why should you vary your spelling?
9. What are four ways of maximizing your search results?
10. What type of site is the .gov domain? .org? .net?
11. What questions should you ask when evaluating Internet resources?

---

http://www.virtualjerusalem.com
Take a virtual tour of the Jewish World from the heart of Israel

Check It Out

Writing Practice

Distance learning, also called distance education, is designed to provide an educational experience from a remote site. You have probably heard of on-line courses while using the Web. But what exactly is distance learning, what are the benefits, and what are the limitations? Are there courses that you could take and receive college credit to graduate sooner than you are currently planning? Using the advanced search features you have learned, research distance learning to find the answers to these questions and any others you might have regarding distance learning. Obtain the **education template** and specific instruction sheet from **http://www.clt.astate.edu/labactivities** to complete this activity.

Lab 6 – Directory Surfing

Learning Objectives

Upon completion of this lab, you will be able to:
- Surf through directories
- Search in searchable directories
- Question directories
- Conduct general and specific searches

Topical Coverage

Upon completion of this lab, you will have explored the following topics:
- Graduate programs in international relations
- Careers
- Health
- International travel
- Foreign languages
- Teen smoking
- Hobbies
- Measurements conversions
- Social Issues
- Various directories
- Continuing education
- Vacation plans

Completion Time

The completion times for each lab activity are shown below. These are only estimated times. Learning the material is much more important than progressing quickly through the activities.

Activity	Completion Time	Activity	Completion Time
Guided Practice 1	46 min	Enrichment 1	23 min
Guided Practice 2	23 min	Enrichment 2	17 min
Guided Practice 3	35 min	Enrichment 3	17 min
Reinforcement 1	17 min	Enrichment 4	29 min
Reinforcement 2	29 min	On-Line Reading	Will vary
Reinforcement 3	Will vary	Writing Practice	Will vary

Lab 6 – Directory Surfing

Guided Practice 1

Directory searching or category searching, also known as, surfing the Web is put to use when you need to find general information about a topic. If you were new to a topic, it would be difficult to conduct a specific search using a search engine. Therefore, this Guided Practice activity will help you to gain general surfing experience to prepare you to conduct more specific searches in the future. Assume you have been given an assignment to choose a graduate program in international relations. This activity will help you practice using Yahoo! as the guide. If you were conducting this research on your own you might have to browse more links, but this exercise will keep you from straying. (Remember that the links could be different than these. Make necessary adjustments when the directions are different).

1. Open **http://www.yahoo.com** The Yahoo! home page will be similar to the one shown in Figure 6-1.

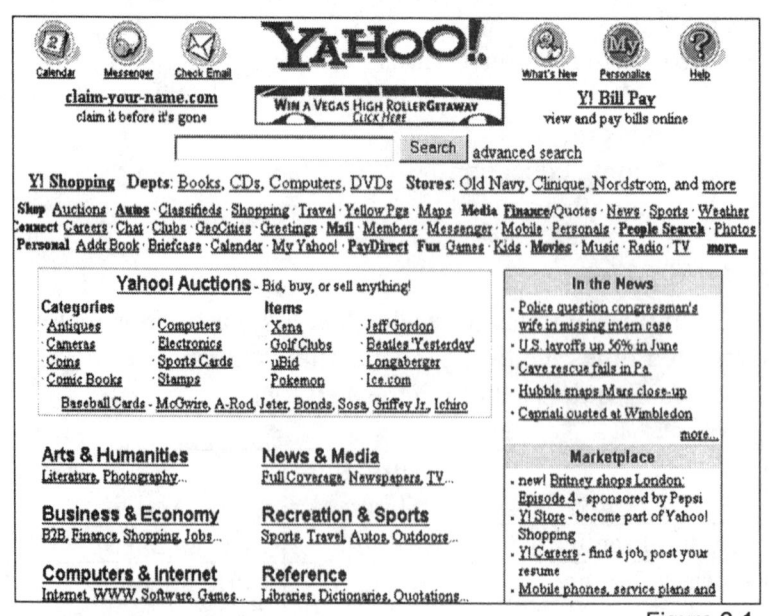

Figure 6-1

2. Click the **College and University** link under the **Education** link.

3. Click the **Graduate Education** link.

4. Key International Relations in the **Search** text box, click the **just this category** button, and then click **Search** as shown in Figure 6-2.

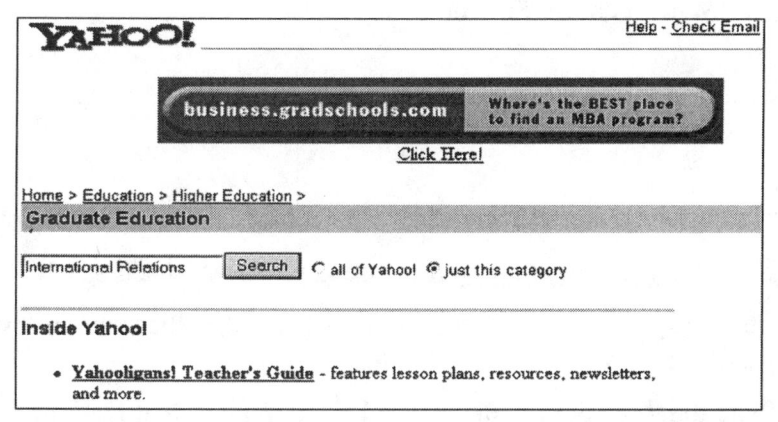

Figure 6-2

5. Now that you are at a page with several document links, click through the links to locate international relations programs for two schools.

6. Print the documents and using a word processor write a comparison between the two schools and their respective programs.

Guided Practice 2

Directories that are combined with search engines are called hybrid search engine directories. Calling them searchable directories shortens this term. This Guided Practice will give you practice in using the Excite searchable directory.

1. Open **http://www.excite.com**

2. Click the **Careers** link located with the **Tools** links. The **Find Your Dream Job** section of the document is shown in Figure 6-3.

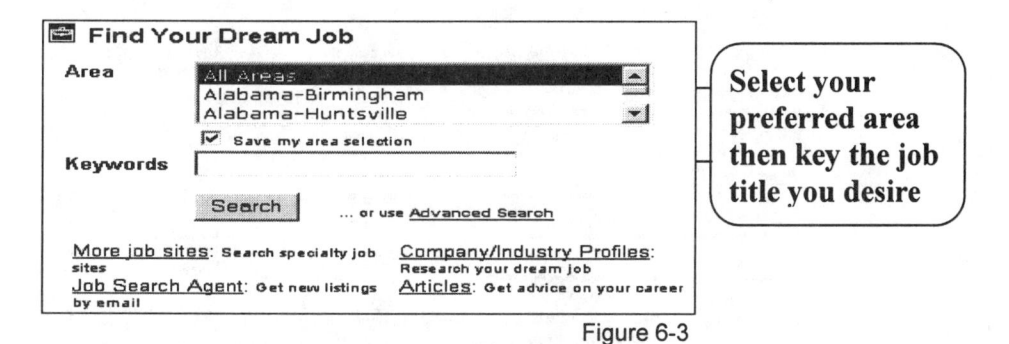

Figure 6-3

3. Choose the area where you want to work.

4. Key the job title you desire in the **Keywords** text. Click **Search**.

1. What location did you choose?	
2. What keyword did you choose?	

5. Click the **use Cool Notify** link and key your email address.

6. Check if you read your email in HTML. Check that you are over 18 years old.

7. Read the three comments marked **Yes!** and deselect if desired. Click **Submit Cool Notify**.

8. Click **Back** to return to the job listing links. Choose a job that interests you and print the advertisement.

9. Only click the **Apply Online to company name** link if you are truly interested in the position.

Guided Practice 3

About.com is a directory whose selling point is that "each site in our unique network is run by a professional Guide who is carefully screened and trained by About." This Guided Practice will give you practice in searching this directory.

1. Open **http://www.about.com** The site index for About-The Human Internet will be similar to that shown in Figure 6-4.

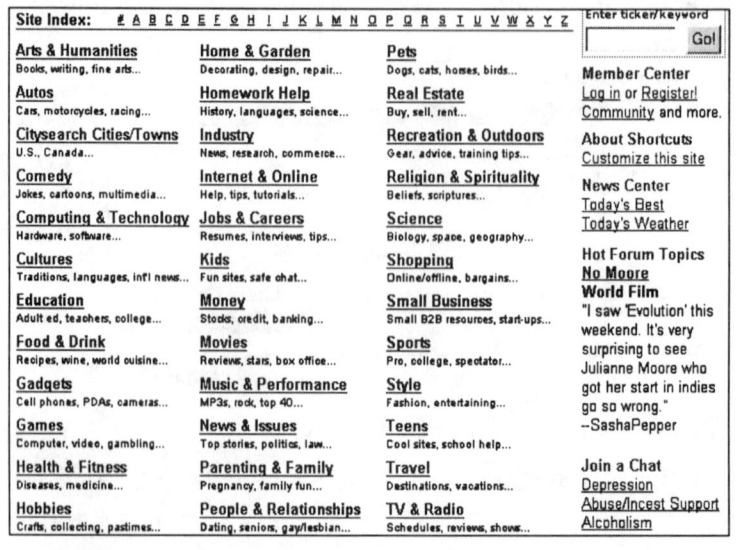

Figure 6-4

2. Click the **Health/Fitness** link.

6-4   Lab 6 – Directory Surfing

3. Click the **Alcoholism** link under the **Recovery/Addiction** link.
4. From the **Subjects** frame on the left, click the **College Drinking** link.
5. Select and read two articles. Print the first page of each article. In the table below write two things you learned from reading the articles about this subject.

1.
2.

6. Conduct two other searches on topics of your choice. Save the documents as text files then open them in a word processor and format them appropriately. Include the citation at the end of the document (refer to Lab 1), then print the documents.

Reinforcement 1

The WWW has made international travel less worrisome with all the information that can be at your fingertips in an instant. One of the first questions you might ask is how is my money converted? This reinforcement activity will show you just that.

1. Open **http://www.xe.net/ucc/**. The Main Menu is shown in Figure 6-5.

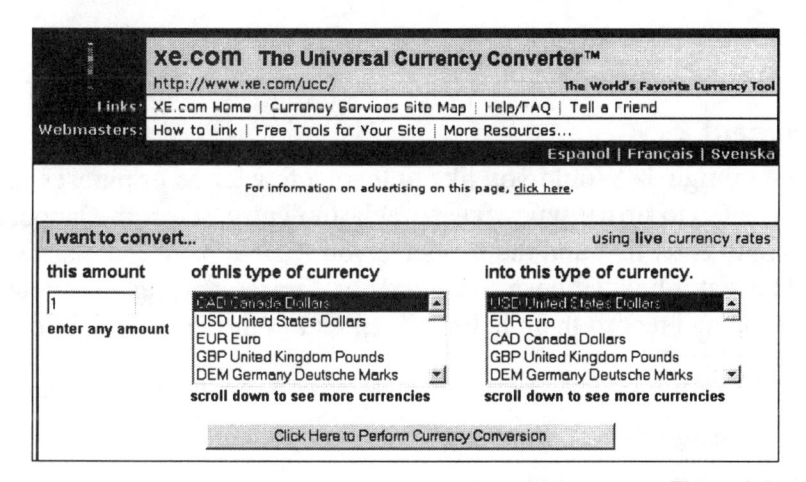

Figure 6-5

2. Choose USD United States Dollars in the **of this type currency** column, then choose BSD Bahamas Dollars in the **into this type of currency** column.

3. Click the **Click Here To Perform Currency Conversion** button, and then print the conversion.

1. What is the conversion for 16 USD?	
2. What is the conversion for 5,000 USD?	

4. Choose two more currencies to convert from and to and record your answers in the table.

Converted From:	**Converted To:**
1. Rate for 18	
2. Rate for 95	
Converted From:	**Converted To:**
1. Rate for 900	
2. Rate for 10,000	

Reinforcement 2

Are you bilingual? Would you like to learn a few terms or phrases in another language? Go to **http://www.freetranslation.com/** Select the language you are using to enter text and the language you wish to have it translate to in the Text Translator box. Choose five words or phrases starting with your name to translate and record them below. Keep text in the Text Translator Edit window. Click Translate!

What language did you translate to?	
Word or Phrase	**Translation**
1.	
2.	

3.	
4.	
5.	

Reinforcement 3

This activity will reinforce the skills for using different directories. Using Yahoo (**http://www.yahoo.com**), then Excite (**http://www.excite.com**), conduct a search to research the effect television has on teen smoking. Print the results pages, then answer the following questions in an email to your instructor.
1. Were the results the directories returned the same or different?
2. How were the results presented in each directory?
3. Which directory was easier to use; which returned more useful results? Explain your answer.

Refer to two of the documents that were returned and correctly cite them in the following table. (Refer to Lab 1 for specific information on citing sources, if necessary).

Citations
1.
2.

Enrichment 1

What kind of surfing do you do? Is it general or specific? When do you do each? Surf in a specific manner on a topic of your choice.

1. What is the focus of your search?	
2. List the beginning link and the last link you followed.	

Now, surf in a general manner until you come to something that catches your attention and you become ready to switch topics.

1. What became the focus of your general search?	
2. List the beginning link and the last link you followed.	

Enrichment 2

Learning more about your hobby is easy with a directory. Go to **http://www.yahoo.com** Click the **Recreation & Sports** link, then click **Hobbies**. Read the information listed about your hobby and print the first page of one useful site. Search within the category to try and locate the history of your hobby. If your search produces a document print the first page. If not, search the entire Web to locate a document and print the first page.

Enrichment 3

Being able to convert weight, capacities, length, area, speed, temperature, pressure, time, etc. quickly and easily can be very useful. And, you do not have to be a scientist to do it, all you need is a site on the Web. Begin your search by going to a directory and performing a Category search under **Science**. Click a link similar to "Measurements Converter," or key the phrase in the Search text box. Choose one or more documents that would allow you to answer the following questions.

1. How many miles are in 327 kilometers?	
2. What is the Fahrenheit measurement for 310.15 degrees Kelvin?	
3. How many square meters are in 12 square feet?	
4. How many nautical miles equal 7 miles?	
5. How many liters equal 6 gallons?	

http://www.jcrew.com
Casual clothes for people on the go.

Check It Out

Enrichment 4

Surf through **http://www.directhit.com** (a partner of Ask Jeeves) to locate items of interest to you. Practice asking Jeeves a question. When you are finished surfing, click the **Society** link, then click the **Issues** link. Click a **Sub-Category** of interest to you. Locate two documents that you could use to write a research paper on the topic with. Crate a reference list based on the Citation Guide in Lab 1. Return to the Sub-Categories and choose a second topic and develop another reference list.

On-Line Reading

Go to **http://www.med.usf.edu/~kmbrown/finding_info.htm** Read through the document titled, *Finding Information on the Web–Directories and Searching*. If the document is no longer at that site, open the document from **http://www.clt.astate.edu/labactivities** Locate the answers to the following questions. Compose and send an email message to your instructor with your answers.

1. What are the characteristics of directories?
2. Why should you become familiar with several directories?
3. How does the Argus Clearinghouse bill itself?
4. What is another name for The McKinley Internet Directory?
5. Who staffs the W3 Virtual Library by subject?
6. If Yahoo retrieves no results what happens?

About the Author: Dr. Kelli R. McCormack-Brown is an Associate Professor in the College of Public Health at the University of South Florida. She is currently investigating the motivators and barriers to using technology among health educators. Dr. McCormack-Brown teaches Foundations of Health Education as both a Web-enhanced and Web-based course.

Writing Practice 1

Continuing education is a must in professional careers today and many employers pay for their employees to attend workshops and schools. Using the skills that you have now acquired about surfing and conducting simple and advanced searches, conduct the necessary research to plan to attend a one-week training workshop. The topic of the training should be within your major concentration or be related to the job you currently have. For instance, if you are an Information Systems major you may want to attend a one-week workshop on learning Visual Basic. Open the **training template** from **http://www.clt.astate.edu/labactivities** for the instructions to be used to write the proposal.

Writing Practice 2

Planning a vacation takes a different course when the World Wide Web enters the picture. With the surfing and searching you have learned you need a vacation. So why not get started by clicking a mouse. Open the **vacation template** at **http://www.clt.astate.edu/labactivities** to obtain the writing instructions for this activity.

> http://www.audubon.org
> The ultimate in bird watching.

Check It Out

Lab 7 – Web Page Design

Learning Objectives

Upon completion of this lab, you will be able to:
- Create a simple Web page
- Download clipart and animation
- Add formatting elements to a Web page
- Understand HTML tags
- Validate HTML code
- Create a personal home page
- Understand the use of color
- Understand programming terms
- Understand Web accessibility

Topical Coverage

Upon completion of this lab, you will have explored the following topics:

- Clipart and animations
- Color psychology
- Java and related terms
- Americans with Disabilities Act
- Authentic Web sites
- Web design mistakes
- Corporate Web sites
- Personal Web sites

Completion Time

The completion times for each lab activity are shown below. These are only estimated times. Learning the material is much more important than progressing quickly through the activities.

Activity	Completion Time	Activity	Completion Time
Guided Practice 1	40 min	Enrichment 1	46 min
Guided Practice 2	29 min	Enrichment 2	29 min
Guided Practice 3	46 min	Enrichment 3	Will vary
Reinforcement 1	23 min	Enrichment 4	Will vary
Reinforcement 2	23 min	On-Line Reading	Will vary
Reinforcement 3	Will vary	Writing Practices	Will vary

Lab 7 – Web Page Design

Guided Practice 1

Creating a simple Web page doesn't require a lot of tools, a text editor such as Notepad and a browser to view the page are all you need. In this Guided Practice activity you will create a simple Web page. You will view the page in your browser as you are building it so you will make a connection between what you are keying (the input) and what the page looks like (the output). The HTML tags that you are using in the Guided Practice activities will be explained in Reinforcement 1. If you would like a better understanding of the tags before you begin, complete Reinforcement 1 now.

1. Open your browser, and then open Notepad found in the Windows Accessories group.

   NOTE: While the directions call for Notepad to be used, any text editor or word processing program may be used.

2. Key the following text in Notepad:

   ```
   <html>
   <head><title></title></head>
   <body>
   </body>
   </html>
   ```

3. Insert your *WWW Lab Activities Data Disk* into the disk drive and choose **Save As** in the **File** menu.

4. Choose **3 ½ Floppy (A:)** from the **Save in** list.

5. Choose **All Files (*.*)** from the **Save as type** list.

6. Key template.htm in the Filename text box, then click **Save**. All Web pages begin with these tags, so you have created a template that you may reuse over and over.

7. Choose **Save As** in the **File** menu again to save the document as **test1.htm** This will create an identical copy of the template that you created and you will not take the chance of making unwanted changes to the template.

8. Edit the document as shown by adding the text that is in boldface type.

   ```
   <html>
   <head><title>My first page</title></head>
   <body>
   ```

    Hi, my name is [enter your name here]
    </body>
    </html>

9. Choose **Save** in the **File** menu to update the file with your changes.

10. Keeping Notepad open, switch to your browser. Choose **Open Page [Open]** in the **File** menu, key a:\test1.htm Click **Open [OK]**.

11. Notice that **My first page** is displayed on the title bar of the browser and the text, **Hi, my name is [your name]** appears in the browser window. Your browser will look similar to the one in Figure 7-1.

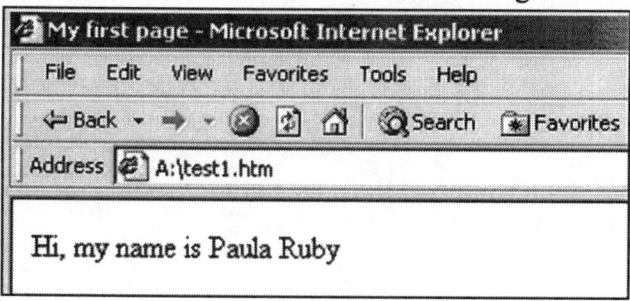

Figure 7-1

12. Switch to Notepad and edit the document by adding the text that is in boldface type.

    <html>
    <head><title>My first page</title></head>
    <body>
    Hi, my name is [your name]
    Hi
    </body>
    </html>

13. After each editorial change you must choose **Save** in the **File** menu to update the file with your changes.

14. Switch back to your browser, and after each editorial change you must click the **Reload [Refresh]** button on the browser to view your changes. You should see the second word Hi, but it is not *below* the first word, Hi, where you might have expected it to be.

15. Switch to Notepad and edit the document by adding the text that is in boldface type.

    <html>
    <head><title>My first page</title></head>
    <body>
    Hi, my name is [your name]**
**
    Hi
    </body>
    </html>

16. Save the changes, then switch back to your browser, and click the **Reload [Refresh]** button. The page should now be similar to the one in Figure 7-2.

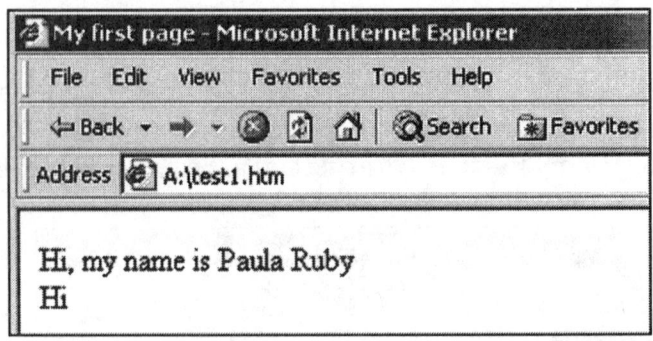

> My first page - Microsoft Internet Explorer
> File   Edit   View   Favorites   Tools   Help
> Back ▼ → ▼ ⊗ ⟳ ⌂ | Search Favorites
> Address A:\test1.htm
>
> Hi, my name is Paula Ruby
> Hi

Figure 7-2

17. Switch to Notepad and edit the document by adding the text that is in boldface type.

    ```
    <html>
    <head><title>My first page</title></head>
    <body>
    Hi, my name is [your name]<br>
    <h2>Hi</h2>
    </body>
    </html>
    ```

18. Save the changes, then switch back to your browser, and click the **Reload [Refresh]** button. You should now see the second word Hi in Headline style 2, which is the second largest headline style.

19. Switch to Notepad and edit the document by adding the text that is in boldface type.

    ```
    <html>
    <head><title>My first page</title></head>
    <body bgcolor="silver">
    Hi, my name is [your name]<br>
    <h2>Hi</h2>
    </body>
    </html>
    ```

20. Save the changes, then switch back to your browser, and click the **Reload [Refresh]** button. You should now see a silver background color.

21. Switch to Notepad and edit the document by adding the text that is in boldface type.

    ```
    <html>
    <head><title>My first page</title></head>
    <body bgcolor="silver" text="blue">
    Hi, my name is [your name]<br>
    <h2>Hi</h2>
    </body>
    </html>
    ```

22. Save the changes, then switch back to your browser, and click the **Reload [Refresh]** button. The text should now be blue.

23. Switch to Notepad and edit the document by adding the text that is in boldface type.

    ```
    <html>
    <head><title>My first page</title></head>
    <body bgcolor="silver" text="blue">
    Hi, my name is [your name]<br>
    <h2>Hi</h2><br>
    <img src="box.gif">
    </body>
    </html>
    ```

24. Save the changes, then switch back to your browser, and click the **Reload [Refresh]** button. You should now see a yellow colored box below the headlined text similar to the one in Figure 7-3.

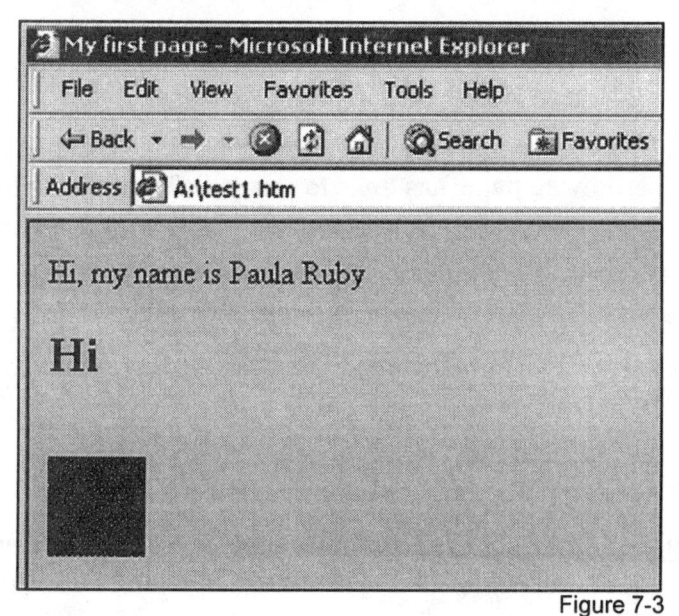

    Figure 7-3

25. Click the **Print** button on the toolbar to print your page.

Guided Practice 2

To add interest to Web pages, oftentimes you will want to include some sort of graphic, like photographs, illustrations, clip art, and animations. Many times you will not have the time or the expertise to produce your own. This is where looking on the World Wide Web can be an asset. During this activity you are going to practice downloading clipart and animations from barrysclipart.com

1. Open your browser, and then open Notepad found in the Windows Accessories group.

   NOTE: While the directions call for Notepad to be used, any text editor or word processing program may be used.

2. Insert your *WWW Lab Activities Data Disk* and open **template.htm** that you created in Guided Practice 1.

3. Choose **Save As** in the **File** menu to save the document as **test2.htm**

4. Edit the document by adding the text that is in boldface type.

   ```
   <html>
   <head><title>My second page</title>
   </head>
   <body>
   Welcome to [your name] page.<br>
   This page demonstrates clipart and animations.
   </body>
   </html>
   ```

5. Choose **Save** in the **File** menu to update the file with your changes.

6. Keeping Notepad open, switch to your browser.

7. Open **http://www.barrysclipart.com** This site, as shown in Figure 7-4 on the following page, has links to many different places to get free clipart, and even has a search feature.

8. Key airplane in the **Find: (e.g.: sports +football)** text box, click **9** for Items: and click **Go** as shown in Figure 7-4 on the following page.

9. Right-click on an airplane image and choose **Save Image As [Save Picture As]**.

10. Save the image to the 3 ½ Floppy (A:) as airplane and the appropriate extension [**airplane.jpg** or **airplane.gif**].

11. Choose a link in the **Cool Animations** section, and then click a topic. Click the download link for the selected animation, then follow steps 9-10 above.

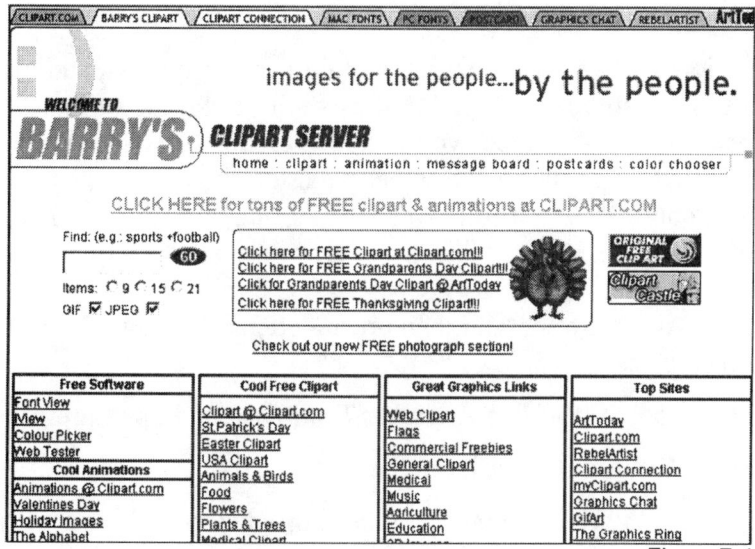

Figure 7-4

12. Switch to the Notepad to edit the document as shown.

    <body>
    Welcome to [your name] page.
    This page demonstrates clipart and animations.
    <img src="filename.ext"> [where filename.ext is the name and extension]
    <img src="filename.ext"> [where filename.ext is the name and extension]
    </body>

13. Click **Save** to save your changes.

14. Keeping Notepad open, switch to your browser. Choose **Open Page [Open]** in the **File** menu, key a:\test2.htm and click **Open [OK]**. Your browser will look similar to the one shown in Figure 7-5.

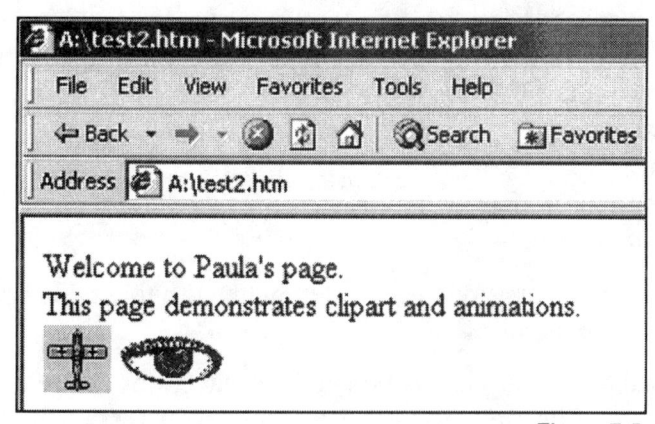

Figure 7-5

15. Click the **Print** button on the toolbar to print your page.

Lab 7 – Web Page Design

Guided Practice 3

In addition to changing background colors, coloring text, and inserting images, there are many ways HTML can display information. This Guided Practice activity introduces you to a few: creating links to other pages you have created, creating links to other pages on the Internet, using background textures, controlling the color of the links, and controlling where text is placed.

1. Open your browser, and then open Notepad found in the Windows Accessories group.

2. Insert your *WWW Lab Activities Data Disk* and open **template.htm**

3. **NOTES:** From this point in the text:

    a. The HTML document will be shortened to include only two lines of the preceding text, the text to be keyed, and two lines of following text.

    b. The directions to save the HTML document after each editing session will be omitted.

    c. The directions to Reload [Refresh] the browser at the beginning of each viewing session will be omitted.

4. You are now going to edit the template.htm document so the template will always include the style sheet. Edit the template by adding the text that is in boldface type.

    <!DOCTYPE HTML PUBLIC "-//W3C//DTD HTML 4.0 Transitional//EN"
    "http://www.w3.org/TR/REC-html40/loose.dtd">
    <html>
    <head><title></title>
    <META HTTP-EQUIV="Content-Type" CONTENT="text/html;
    charset=ISO-8859-1">
    </head>
    <body>
    </body>
    </html>

5. Because you are updating the template with these changes, choose **Save** in the **File** menu.

6. Now that the template is updated and saved for future use, choose **Save As** in the **File** menu to save the document as **test3.htm**

7. Edit the document by adding the text that is in boldface type.

    "http://www.w3.org/TR/REC-html40/loose.dtd">
    <html>
    <head><title>**My third page**</title>

```
<META HTTP-EQUIV="Content-Type" CONTENT="text/html;
charset=ISO-8859-1">
</head>
<body>
```
Welcome to [your name] page
```
</body>
</html>
```

8. Keeping Notepad open, switch to your browser. Choose **Open Page [Open]** in the **File** menu, key a:\test3.htm and click **Open [OK]**.

9. Notice that **My third page** is displayed on the title bar of the browser and you should see the words **Welcome to [your name] page** in the browser window.

10. Edit the document by adding the text that is in boldface type.

    charset=ISO-8859-1">
    </head>
    <!--Modify the body to display a texture, control the color of the text, control the color of links you have not visited, and control the color of the links that have been visited-->
    <body
    background="texture.gif"
    text="#0000ff"
    link="#ff0000"
    vlink="00ff00">
    <H3>Welcome to [your name] page**</h3>**

11. View the changes in your browser.

12. Edit the document by adding the text that is in boldface type.

    vlink="00ff00">
    <H3>Welcome to [your name] page</h3>
    <!--Create a link to your test1.htm-->
    **<a href="test1.htm">Click to go to my 1st page</a>
**
    <!--Create a link to test2.htm, using an image-->
    **<a href="test2.htm"><img src="2nd.gif" alt="2nd"></a>
**
    <!--Create a link to AltaVista-->
    **<a href="http://www.altavista.com">AltaVista</a>
**
    <!--Align some elements to be centered-->
    ** I'll be adding more soon
**
    **<img src="bye.gif" alt="bye"> **
    </body>
    </html>

13. View the changes in your browser. The final document will be similar to the one in Figure 7-6 on the following page.

> **Welcome to Paula's page**
>
> Click to go to my 1st page
>
> **2nd**
>
> AltaVista
>
> I'll be adding more soon
>
> *Thanks for coming bye!*

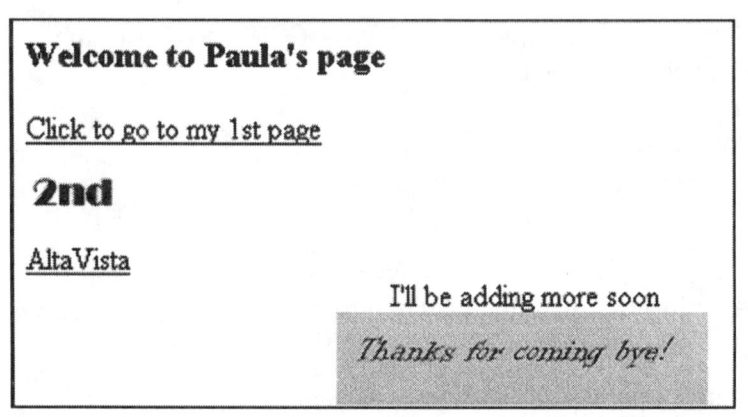

Figure 7-6

14. Test to see that the links work by clicking on them and then clicking the **Back** button on the toolbar to return to your page.

15. Click the **Print** button on the toolbar to print your page.

Reinforcement 1

Now that you have practiced creating and editing some HTML documents and have viewed the end results, it would be beneficial to locate explanations of all the coding and tags that were needed to create the documents. Open the **html template** from **http://www.clt.astate.edu/labactivities** and referring to the following Web site, write the use for each element used in the documents you created. The Web site to use is **http://www.ncsa.uiuc.edu/General/Internet/WWW/HTMLPrimerP1.html**

Reinforcement 2

As you have worked through some of these exercises, you have undoubtedly misspelled tags, forgotten quotation marks, or not placed an ending tag where it was required. It may have taken you a long time to find the error, and what's worse, there may still be errors in your page that don't display as a problem, or perhaps your browser has corrected for you. The only way to make sure you have entered your code correctly is to get it validated. That is, have your code checked by a program that understands HTML, and can spot errors. It is still up to you to correct the errors, but it is a great feeling when you get the message, "Congratulations, no errors!"

1. Open your browser, and then open Notepad found in the Windows Accessories group.

2. Insert your *WWW Lab Activities Data Disk* and open **test4.htm**

3. Keeping Notepad open, switch to your browser.

4. Open **http://www.htmlhelp.com/tools/validator/**

5. Click the **validate files on your computer** link.

6. Key a:\test4.htm in the text box as shown in Figure 7-7.

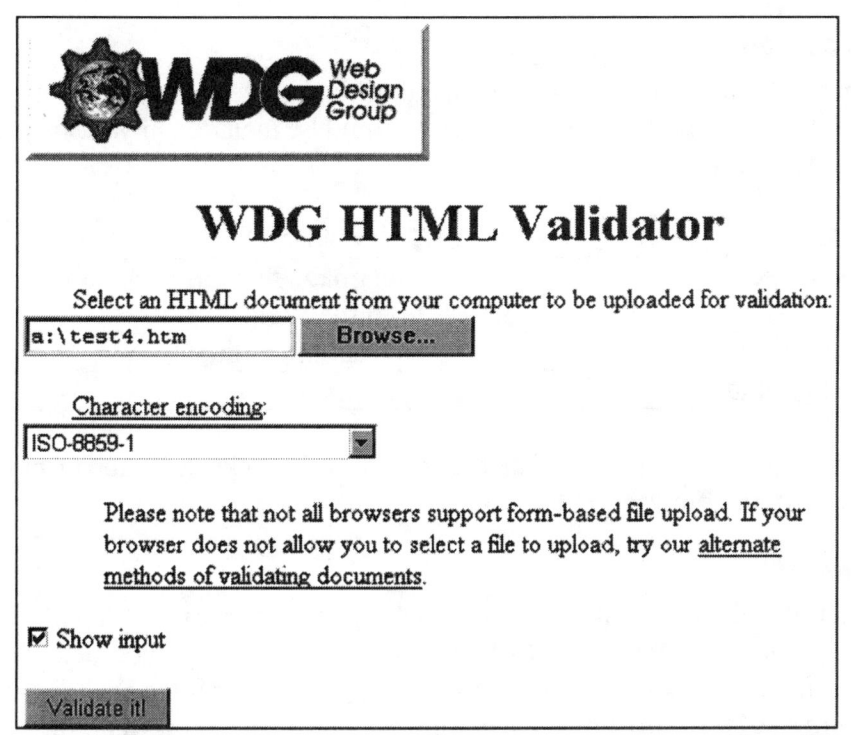

Figure 7-7

7. Click **Validate it!**
8. Click **Print** on the toolbar to print the document.
9. Compare each error to the test4.htm document that is open in Notepad. Make the necessary corrections.

   NOTE: Some errors will go away automatically when you fix the others.

10. When you finish correcting the errors, choose **Save** in the **File** menu to update the document.
11. Switch to your browser, then press **Reload [Refresh]** to restart the validation process.
12. Repeat the process until all the errors are fixed and you receive the "Congratulations, no errors!" message. While you are correcting your errors, or when you are finished, return to the document and replace your name with the [your name] text on or about Line 17.
13. Click **Print** on the Toolbar to print the Congratulations message.
14. View the changes in your browser.
15. Test to see that the links work by clicking on them and then clicking the **Back** button on the Toolbar to return to your page.

16. Click the **Print** button on the Toolbar to print your page.

Reinforcement 3

Using the skills that you have learned thus far, create a personal home page using HTML. The following elements should be included in your page and are the minimum requirement.

Title	Text color
Body Text	One image that has been downloaded from the Web
Two levels of headings	One comment line
Background color	Two links

When you are finished, validate the code, test the links, and then print the page from the browser.

Enrichment 1

Selecting colors for your Web site is an important part of the planning and creation of your documents. Because you like a color does not mean that it will elicit the response you want from your page. Search the Web for a minimum of three documents pertaining to color psychology and the Web. Answer the questions in the table record the three URLs that you used to locate your answers.

1. What is color psychology?
2. What is a color-safe palette?
3. What does the color cyan represent in finance? In the medical field?
4. What does the color brown represent? Black?
5. In Western cultures what does the color white represent? In China?
URLs Used
1.
2.
3.

Enrichment 2

Many similar terms abound when using the Web and many of these terms sound like they are related to or are part of the Java programming language. But how are they related, or are they related at all? Conduct a search to define the terms below to better understand the relationship to Java.

Term	Definition
1. Java	
2. Java Applets	
3. Java Servlets	
4. JavaBeans	
5. JavaScript	
6. Jscript	
7. Vbscript	

Enrichment 3

As the use of the WWW continues to expand, issues continue to be raised and questions continue to be asked about who has the right to use the Web. One issue is in regard to the Americans with Disabilities Act (ADA). Specifically, are Internet Web pages required to be accessible to people with certain disabilities? Or are Web pages in compliance with the ADA? And is there a way to check your Web pages for accessibility problems? Using the advanced searching skills you have learned, research this topic and using at least three references, create a word processing document to compose your answer in paragraph format. At the end of the document, key and center the word, References, then create a complete citation for each Web document you used. (Refer to Lab 1 for creating citations).

Enrichment 4

The introduction of the Web has brought an onslaught of authors–some authentic and some not. Be aware that not all of the resources you find on the Web are legitimate. When evaluating Internet resources ask the following questions:

1. Are editors listed?
2. Is the document free of typographical and grammatical errors?
3. Is the date of the last update current?
4. Is the document from a well-known source?
5. Are facts, quotes, and statistics referenced?

Using the search engine and topic of your choice locate and evaluate two corporate Web sites based on the previous questions. Search for the sites using two additional search engines. Next, repeat the process for two Web sites that you find questionable. Create a word processing document to record your findings. For each site:

1. Name the first search engine used.
2. Name the two additional search engines and state if the Web sites were located.
3. Answer the five questions listed above.
4. Write a brief assessment of the site.

On-Line Reading

Go to **http://www.useit.com/alertbox/990530.html** Read through the document titled, *The Top Ten New Mistakes of Web Design*. If the document is no longer at that site, open the document from *the WWW Lab Activities Data Disk*. Locate the answers to the following questions. Compose and send an email message to your instructor with your answers.

1. What is Mistake #1?
   a. What three sins break **Back**?
2. What is Mistake #2?
   a. Why do designers open new browser windows?
3. What is Mistake #3?
   a. What is one of the most powerful usability principles?
   b. What are GUI widgets?
4. What is Mistake #4?
   a. Why should you include a biography on a Web site?
5. What is Mistake #5?
   a. What is the added percentage increase of cost for keeping archives?
   b. What is linkrot?
6. What is Mistake #6?
   a. What happens when a page moves?
7. What is Mistake #7?
   a. What are the two things headlines do?
8. What is Mistake #8?
   a. What are the benefits of using buzzwords?
9. What is Mistake #9?
   a. What should you do to increase response times?
10. What is Mistake #10?
    a. What are three design problems?

---

http://www.the-light.com/colclick.html
A page that shows hexadecimal numerical values that can be substituted for word colors.

Check It Out

About the Author: Jakob Nielsen, Ph.D., is a User Advocate and principal of the Nielsen Norman Group, which he co-founded with Donald A. Norman (former VP of research at Apple Computer). Until 1998 he was a Sun Microsystems Distinguished Engineer and the company's Web usability guru. Dr. Nielsen coined the term "discount usability engineering" and has invented several usability techniques for fast and cheap improvements of user interfaces, including heuristic evaluation. He holds 33 United States patents, mainly on ways of making the Internet easier to use.

Writing Practice 1

Many companies today feel the need to quickly develop a presence on the Web. Because of this employees who are experienced users of the Web (or are vaguely familiar with the Web) are often required to design and create the company's Web site, oftentimes without receiving formal training. Assume that you are the employee who has been directed to create the company Web site. Examine and critically evaluate three professionally developed Web pages in the area in which you plan to have a career. For instance, if you are planning to work in delivery services, then you might examine the Federal Express or UPS Web sites. You must develop a matrix for writing a proposal that must be approved before you begin designing the site. Open the **web design template** from **http://www.clt.astate.edu/ labactivities** for the instructions to be used in developing the matrix and writing the proposal.

Writing Practice 2

As you have seen throughout the time you have been using the WWW, Web sites have different designs and content based on the purpose of the site. In Reinforcement 3 you created a home page using HTML code. Web sites are available to lead you through the steps of designing and creating Web site for your intended purposes. Go to **http://www.homestead.com** Choose to set up a site for either your personal home page, a club or organization to which you belong, or one for your hobby. Explore the site including looking at examples of member sites. Also, click on the link: Learning Center and follow the directions there. Create at a minimum a home page for your site using this Web site as a guide.

> http://www.css.nu/
> The Cascading Style
> Sheet Pointers Group

Check It Out

Lab 8 – Legal, Societal Issues, and Government Sites

Learning Objectives

Upon completion of this lab, you will be able to:
- Locate government information
- Search foreign government sites
- Search for banned books
- Search the Library of Congress
- Locate statistical information
- Search for watchdog groups
- Search the Bureau of Labor Statistics

Topical Coverage

Upon completion of this lab, you will have explored the following topics:
- Electronic Privacy Information Center
- Consumer Information Center
- Foreign national symbols and top officials
- Censorship
- Holdings of the Library of Congress
- Thomas and congressional records
- State statistics
- Politics
- Virus myths
- Child pornography
- Occupational Outlook Handbook
- Copyright laws
- Blue Ribbon Campaign for on-line free speech
- Web-bugs

Completion Time

The completion times for each lab activity are shown below. These are only estimated times. Learning the material is much more important than progressing quickly through the activities.

Activity	Completion Time	Activity	Completion Time
Guided Practice 1	23 min	Enrichment 1	Will vary
Guided Practice 2	29 min	Enrichment 2	Will vary
Guided Practice 3	29 min	Enrichment 3	29 min
Reinforcement 1	23 min	Enrichment 4	29 min
Reinforcement 2	35 min	On-Line Readings	35 min
Reinforcement 3	40 min	Writing Practice	Will vary

Lab 8 – Legal, Societal Issues, and Government Sites

Guided Practice 1

EPIC is the acronym for the Electronic Privacy Information Center based in Washington, D.C. It was established in 1994 "to focus public attention on merging civil liberties issues and to protect privacy, the First Amendment, and Constitutional values." This Guided Practice activity will help you become familiar with their Web site.

1. Open **http://www.epic.org**
2. Click the **About EPIC** link and read the information.
3. Click the **Privacy** link. Record three hot topics and new resources in the following table.

1.
2.
3.

4. Click the **Back** button on the Toolbar until you return to the EPIC home page.
5. Read the abstracts of the latest news, and then pick one story on which to place your focus. Each story will have one or more links to follow.
6. Click one of the links and print only Page 1 of the document.
7. Create a **Bookmark** or mark this site as a **Favorite**.

Guided Practice 2

This Guided Practice activity will lead you through locating information from the Consumer Information Center Web site and will direct you in printing booklets that are provided free on the site.

1. Open **http://www.pueblo.gsa.gov** The Consumer Information Center Web site will be similar to the one in Figure 8-1.

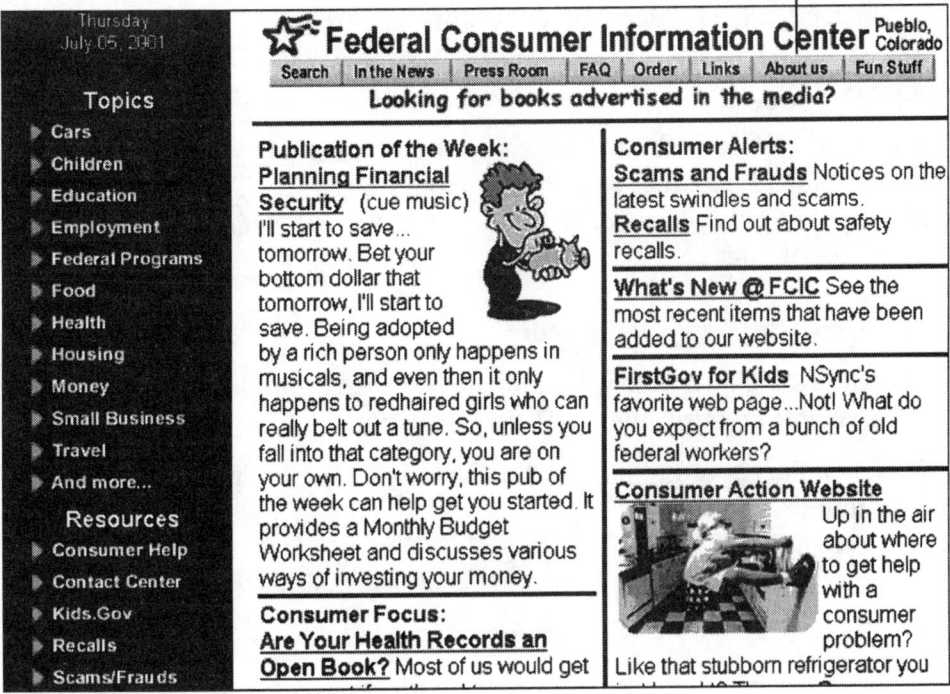

Figure 8-1

2. Click the **About Us** link at the top of the page and read the background information about the CIC.

3. Click the links in the left frame looking for publications of interest to you. Scroll through the list of links. When you find one that interests you, click on that link.

4. Read the information and print only Page 1 of the booklet.

5. Choose two more booklets from two different topics and print Page 1 of each booklet. Record the topics and booklet titles that you chose in the following table.

Topic	Booklet Title
1.	
2.	
3.	

6. If you would like to have the booklet mailed to you and/or want a booklet where a fee is required, then you will need to place an order. To do this, click the **Back** button on the Toolbar to return to the home page, then click the **Order** link and follow the directions.

Guided Practice 3

The World Wide Web has opened up the world to us by bringing information about other countries into our daily lives. This Guided Practice activity will give you practice in linking to Australia and then other countries of your choice.

1. Open **http://www.yahoo.com**
2. Click the **Government** link, then click the **Countries** link. A list of available countries will now be displayed.
3. Click the **Canada** link and the display will be similar to the one in Figure 8-2.

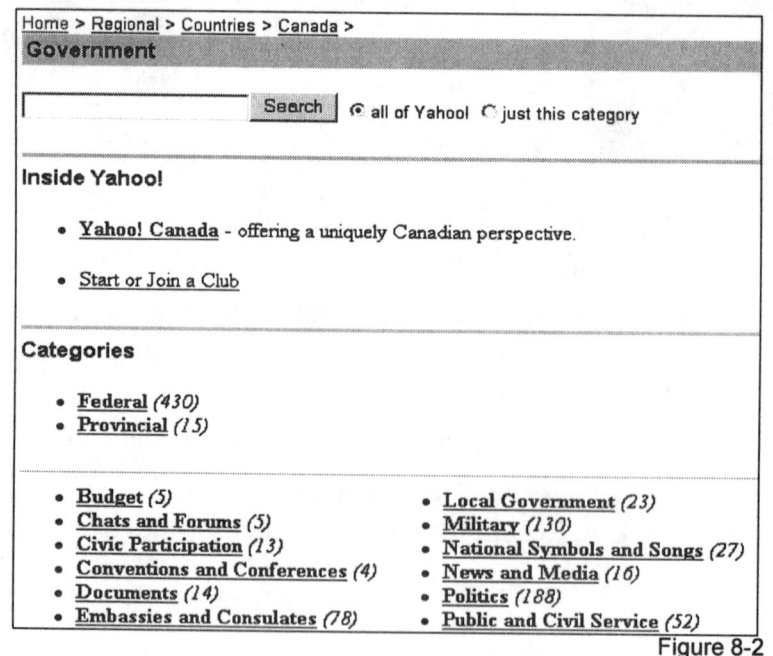

```
Home > Regional > Countries > Canada >
Government

[        Search  ] ⊙ all of Yahoo!  ○ just this category

Inside Yahoo!
  • Yahoo! Canada - offering a uniquely Canadian perspective.
  • Start or Join a Club

Categories
  • Federal (430)
  • Provincial (15)

  • Budget (5)                              • Local Government (23)
  • Chats and Forums (5)                    • Military (130)
  • Civic Participation (13)                • National Symbols and Songs (27)
  • Conventions and Conferences (4)         • News and Media (16)
  • Documents (14)                          • Politics (188)
  • Embassies and Consulates (78)           • Public and Civil Service (52)
```
Figure 8-2

4. Click the **National Symbols and Songs** link, and then click the **National Anthems** link.
5. Click **National Anthem: O Canada** and save the National Anthem as a text document.
6. Click the **Back** button on your browser twice.
7. Click the **National Flags** link.

8-4   Lab 8 – Legal, Societal Issues, and Government Sites

8. Click the **Canadian Clip Art Gallery** link. Select one of the flag graphics from the left frame and save the clip art as a graphic.

9. Format the Anthem and the Flag in a word processing document, then print the document.

10. Locate the National Anthem and National Flag for two other countries and save them to a word processing document as well.

11. Search for the top elected official of each country. Record your answers in the following table.

Country	Top Elected Official	Their Title
1. Canada		
2.		
3.		

Reinforcement 1

Throughout history, books have been censored and/or banned depending on the state of society at the time. The Banned Books On-Line Web site explains the reasons for the censorship/banning for many books that have been affected over the years. The text of each book is also available to read on-line or to print and read later. Many of these books are since out of print and unavailable for purchase or use.

1. Open **http://digital.library.upenn.edu/books**
2. Click on the **Banned Books On-Line** link.
3. Read the discussion surrounding two titles that were "Suppressed or Censored by Legal Authorities."

1. What two titles did you read about?	

4. Read the discussion surrounding two titles that were "unfit for schools and minors."

1. What two titles did you read about?	

5. Create a word processing document to discuss why you think each of these books should or should not have been banned. Given society today would these books be acceptable for reading or would they still be banned. State your remarks in paragraph format. At the end of the document, key and center the word, References, then create a complete citation for each Web document you used. (Refer to Lab 1 for creating citations).

Reinforcement 2

Until the Web, the Library of Congress was something that most people only heard of, but weren't sure what the Library contained or what its purpose was. Now, even though its physical location remains in Washington, D.C. you can gain access through their Web site. This Reinforcement activity will give you the opportunity to surf through the Library, and then locate congressional bills regarding Internet activities.

1. Open **http://www.loc.gov/**
2. Surf through the Web site to see what is available at the Library. While you are surfing, locate the names of the three buildings that make up the Library and record those names in the table on the following page.

1.	
2.	
3.	

3. At the Thomas page, search to locate when Thomas went on-line and for what purpose. Record your findings in the table that follows.

1. Date	
2. Purpose	

> http://www.sina.com
> A Chinese search engine that can be converted to English.

Check It Out

4. Return to the Thomas home page that will be similar to the one in Figure 8-3.

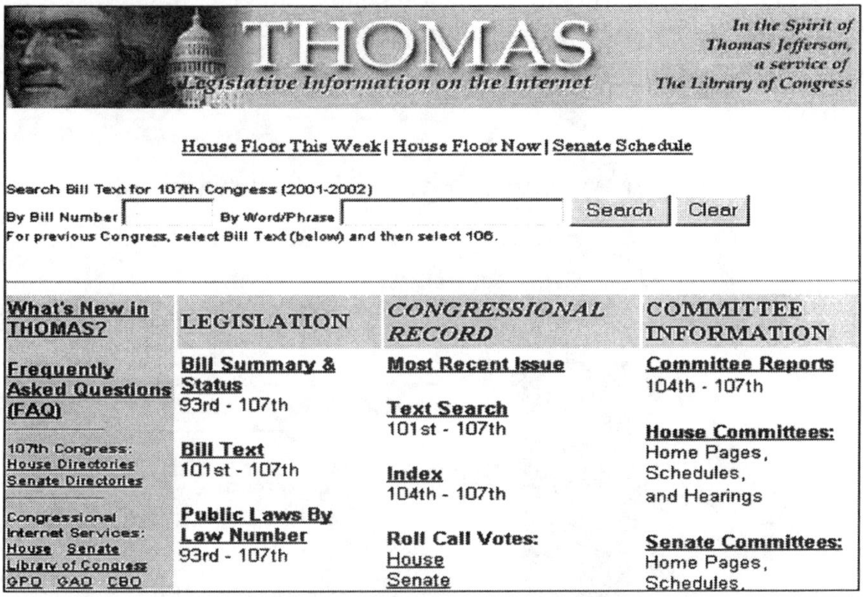

Figure 8-3

5. Key "gambling" AND "internet" in the **By Word/Phrase** text box, then click **Search**.

1. How many bills from Congress were located?	

6. Click on one of the bills and print the Table of Contents page.

7. Conduct three more searches on any controversial Internet topic and list the results.

	Search Term/Phrase	Number of Bills Located
1.		
2.		
3.		

http://www.gorp.com/gorp/activity/hiking.htm
Stop daydreaming and start hiking.

Check It Out

Lab 8 – Legal, Societal Issues, and Government Sites

Reinforcement 3

The U.S. Census Bureau has a very useful Web site that can serve as a directory/search engine in and of itself. During this Reinforcement activity you will find statistical information that pertains to the state in which you live.

1. Open **http://www.census.gov**. The Web site will be similar to the one in Figure 8-4.

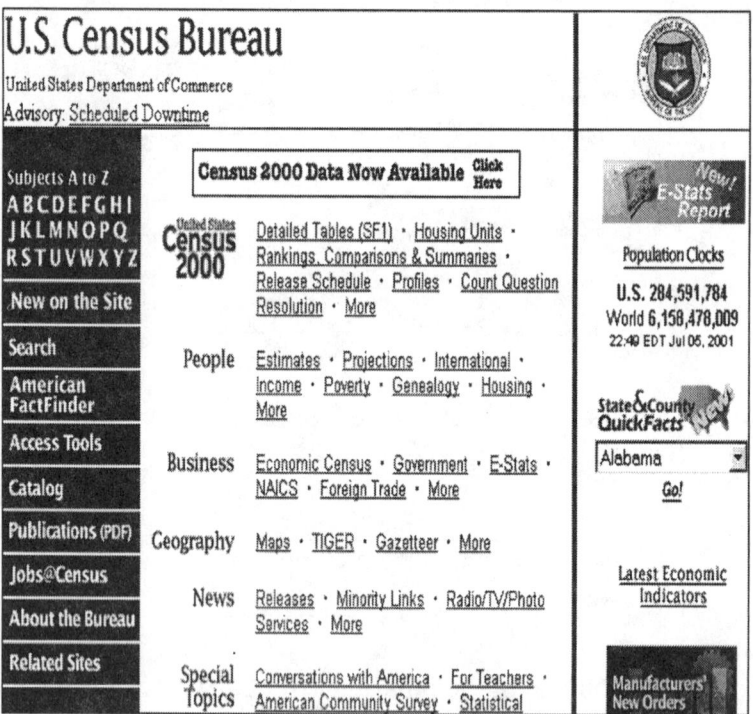

Figure 8-4

2. Using the searching and surfing skills you have learned, answer the questions on the following page for the state in which you live that can be found at the U.S. Census Bureau site.

1. What state did you search?	
2. What is the per capita income?	
3. What is the population and rank compared to the United States?	

4. What is the percentage birth rate to teenage mothers and rank compared to the United States?	
5. What is the rate of violent crime and rank compared to the United States?	
6. What is the median annual pay and rank compared to the United States?	

Enrichment 1

National, state, and local politics have moved to the World Wide Web. No longer do you have to watch the evening news or read the newspaper to learn what is happening. Locate and print the home pages of two members of either the Congress of the United States or the elected officials of your state. Create a word processing document explaining your impressions of the sites. State your remarks in paragraph format. At the end of the document, key and center the word, References, then create a complete citation for each home page viewed. (Refer to Lab 1 for creating citations).

Enrichment 2

More and more the news is reporting accounts of computer viruses infecting the computers and networks around the globe. As well, you may receive email warning you of computer viruses that really aren't viruses at all but are myths or hoaxes. Open the Computer Virus Myths home page at **http://vmyths.com/** Choose three myths and in a word processing document discuss how each one could have garnered attention, what could be done to avoid the myth, and how you would be able to discover that it was a hoax.

> http://www.foodtv.com
> Get recipes from the
> Food Network.
> **Check It Out**

Enrichment 3

Child pornography on the Web is becoming a quickly growing problem. Organizations, called watchdog groups, are forming to combat this travesty. One such organization is The Dallas Association for Decency (DAD). Search the Web for three other watchdog groups to protect our children. Create a word processing document that explains the purpose of each group. At the end of the document, key and center the word, References, then create a complete citation for each home page viewed. (Refer to Lab 1 for creating citations).

Enrichment 4

The future calls and you need to discover what the employment outlook is for someone going into your future profession. Open the Web site for the Occupational Outlook Handbook provided by the Bureau of Labor Statistics at **http://stats.bls.gov/ocohome.htm** Conduct a search for your selected occupation, or one that comes closest to your future ambition and answer the following questions.

1. At what figure do starting salaries begin?	
2. What is the outlook for growth in this occupation?	
3. What prerequisites are needed to enter this field?	

On-Line Reading 1

Go to **http://whatiscopyright.org/** Read through the document titled, *What is Copyright Protection?* If the document is no longer at that site, open the document from *the WWW Lab Activities Data Disk*. Locate the answers to the following questions. Compose and send an email message to your instructor with your answers.

1. What is copyright?
2. When does copyright protection begin and what is required?
3. Is it legal to place the copyright © symbol on your work before it is registered?
4. What is the proper way to place a copyright notice?
5. What are the three reasons information from the Internet can be copied?
6. What does the copyright © symbol at the bottom of a home page apply to?
7. What is fair use?

About the Author: Rebeca Delgado-Martinez V. practices general corporate, business and commercial law, and has worked five years as an associate attorney for major law firms and two years for a telecommunications company. Ms. Delgado-Martinez V. currently resides in France.

---

http://www.realchange.org/
A view of the politicians' closets.

Check It Out

On-Line Reading 2

Go to **http://seattletimes.nwsource.com/** In the Search Web Archive, search for the document titled, *Web bugs nibbling at computer privacy*, and then read through the document. If the document is no longer at that site, open the document from your *WWW Lab Activities Data Disk*. Locate the answers to the following questions. Compose and send an email message to your instructor with your answers, or key your answers in the correct file on your *WWW Lab Activities Data Disk*.

1. What is a Web bug?
2. What are Web bugs known as in the trade?
3. What is the FTC going to do about Web bugs?
4. What happens if one cookie is set?
5. Why do Janssen Pharmaceutical Products deploy Web bugs?
6. How do Web bugs work with the help of a cookie?
7. What do Web bugs represent to consumer watchdogs?

About the Author: This article was published in the *Seattle Times* newspaper. It was written by Robert O'Harrow, Jr. of *The Washington Post*.

Writing Practice

The "Blue Ribbon Campaign" for on-line free speech is sponsored in part by the Electronic Frontier Foundation (EFF). Conduct a search of the Web to research this topic. Write a position paper, for or against, the campaign. Obtain the **blue ribbon template** from **http://www.clt.astate.edu/ labactivities** for the writing instructions.

> http://www.privacy.net
> Analyze your Internet privacy at the Consumer Information Organization.
>
> Check It Out

Index

Site Index

A

www.about.com
www.albion.com/netiquette/
    introduction.html
www.alltheweb.com
www.altavista.com
www.audubon.org

B

www.barrysclipart.com
www.blackvoices.com
www.bloomin.com
www.bluemountain.com

C

www.catalogsavings.com
www.census.gov
www.clt.astate.edu/labactivities
www.collegegrad.com
www.country.com
www.css.nu/

D

www.dawn.com
www0.Delphi.com/navnet/faq/
    history.html
digital.library.upenn.edu/books
www.directhit.com

E

www.eiu.edu/~alsiglam/
www.epic.org
www.excite.com

F

www.financialmarketquotes.com
www.foodtv.com

www.freetranslation.com

G

www.gamesville.com
www.go.com
www.golfacademy.com
www.google.com

H

www.harmony-central.com
www.historychannel.com
www.homestead.com
www.hotmail.com
www.htmlhelp.com/tools/validator/

J

www.jcrew.com

L

www.loc.gov
www.looksmart.com
www.lycos.com

M

www.mapquest.com
www.med.usf.edu/~kmbrown/
    finding_info.htm
www.microimg.com/postcards/
www.moviefone.com
www.msnbc.com
www.mysterynet.com

N

www.nasdaq.com
www.nbci.com

www.ncsa.uiuc.edu/General/Internet/WWW/HTMLPrimerP1.html
www.netforbeginners.about.com
www.netscape.com
www.nyse.com

P

www.philb.com/compare.htm
www.phobe.com/furby/
www.piaget.org
www.privacy.net
www.profusion.com
www.promo.net/pg
www.pueblo.gsa.gov

Q

www.quicken.com/retirement/

R

www.realchange.org
www.rice.edu/Fondren/Netguides/strategies.html
www.ricksteves.com

S

www.safeshopping.org/security/main.html
www.search.com
www.seattletimes.nwsource.com/
www.sina.com
www.sportsillustrated.cnn.com
stats.bls.gov/ocohome.htm
www.stonepages.com/England/englandmain.html

T

www.the-light.com/colclick.html
www.tucows.com

U

www.useit.com/alertbox/990530.html

V

www.virtualjerusalem.com

www.vmyths.com

W

www.webhelp.com
www.webteacher.org
whatiscopyright.org
www.whitehouse.gov

X

www.xe.net/ucc

Y

www.yahoo.com

Z

www.zdnet.com/anchordesk/stories/story/0,10738,2354132,00.html

Text Index

A

About home page, 6-4
Absolutely Amazing Greeting Cards home page, 1-7
Address bar, 1-4
Addresses, 2-7
Academic programs, 5-10, 6-2
Alcoholism, 6-5
AltaVista home page, 2-10, 3-2, 5-2
Alternative medicine, 3-9
Americans with Disabilities Act, 7-13
Animation, 7-6
APA 1-9
Ask Jeeves home page, 5-7
Attachments, 4-5, 4-9
Automobiles, 3-10

B

Bank notes, 5-2
Banned Books On-Line home page, 8-5

Barry's Clipart Server home page, 7-6
Baseball, 5-10
Black Voices home page, 3-9
Bloomin home page, 1-3
Blue Ribbon Campaign, 8-11
Blue Mountain Arts home page, 1-7
Bookmarks, 1-7
Boolean expressions, 5-2, 5-3
Bradley, Phil, 3-10
Bureau of Labor Statistics, 8-10
Bush, George W., 5-5
Business, 4-14

C

Canada, 8-4
Careers, 6-3
Category searching, 6-2
Chat, 2-5, 2-11
Child pornography, 8-9
Chinese search engine home page, 8-6
Citation guide, 1-8
Clipart, 7-5
C|NET home page, 3-3, 5-7
Collegegrad home page, 1-2
Color psychology, 7-12
Computer Virus Myths home page, 8-9
Consumer Information Center home page, 8-2
Continuing education, 6-9
Copyright protection, 8-10
Country home page, 4-4
Cracker Barrel, 2-9
Crystal meth, 5-8
Currency converter, 6-5

D

Dali, Salvador, 3-2
Dallas Association for Decency, 8-9
Delgado-Martinez V., Rebeca, 8-10
Direct Hit home page, 6-9
Directory searching, 6-2, 6-7

Distance learning, 5-11
Distribution list, 4-11
Dogs, 3-7
Domain searching, 5-5, 5-6
Driving instructions, 2-10

E

Electronic mail, 4-2
Electronic Frontier Foundation, 8-11
Electronic Privacy Information Center, 8-2
Electronic resources, 5-8
Emoticons, 4-12
EPIC home page, 8-2
Etexts, 4-8, 4-11
Ethics, 3-9, 5-6, 5-8
Excite home page, 2-11, 5-4, 6-3, 6-7

F

Favorites, 1-7
File attachments, 4-5, 4-10
File Transfer Protocol, 4-6, 4-11
Financial Markets Quotes home page, 2-10
Find people, 2-7
Food TV home page, 8-9
Furby Autopsy, 1-9

G

Gambling, 8-7
Gamesville home page, 3-4
Gates, Bill, 4-14
Genealogy, 3-4
Go Network home page, 3-8
Google home page, 3-6
Gore, Al, 5-5
Gottesman, Ben Z., 2-12
Greeting cards, 1-7

H

Harmony Central home page, 5-3
Hexadecimal values, 7-14
History, 1-8, 1-11, 3-11, 8-5
Hobbies, 3-4, 6-8

Home Page, 7-12, 7-15
Homestead home page, 7-15
Hotmail home page, 4-2
HTML, 7-2, 7-10
HTML Validator, 7-10

I

Illegal drugs, 5-8
Internet Service Provider, 4-14

J

J. Crew home page, 6-8
Java, 7-13
Jewelers, 2-9
Job search, 1-2, 6-3
Jobs, 7-15, 8-10

K

Kennedy, John F., 5-3

L

Language translator, 6-6
Library of Congress home page, 8-6
Location bar, 1-4
Looksmart home page, 2-7
Lycos home page, 3-8

M

Mailing lists, 4-11
Matisse, Henri, 3-8
McCormack-Brown, Kelli R., 6-9
Measurements converter, 6-8
Media, 1-10
MLA, 1-9
Movie stars, 3-10
Moviefone home page, 2-13
MSNBC, 1-9
Music, 3-10
Mystery Net home page, 3-11

N

NASA home page, 4-13
Nasdaq home page, 1-10
NBCi home page, 2-2, 2-4

Netiquette, 4-12
Net for Beginners home page, 4-11
Netscape home page, 2-10
New York Stock Exchange home page, 1-10
News, 1-9, 2-4
Nielsen, Jakob, 7-15
Notepad, 7-2

O

Occupational Outlook Handbook home page, 8-10
O'Harrow, Robert, 8-11
On-line forms, 1-3
On-line free speech, 8-11
On-line shopping, 2-11
Open dialog box, 1-2
Open Page dialog box, 1-2

P

Page Setup dialog box, 1-6
Pakistan news, 1-8
Personal home page, 7-12
Piaget, Jean, 1-4
Picasso, Pablo, 3-8
Portal, 2-2, 2-4, 2-11, 2-12, 2-13
Presidential libraries, 3-8
Politics, 5-5, 8-9, 8-10
Print, 1-4
Privacy home page, 8-11
ProFusion home page, 3-4, 5-7
Project Gutenberg, 4-7, 4-11

Q

Quotations, 5-7

R

Real Change home page, 8-10
Rick Steves home page, 5-8

S

Safe Shopping home page, 4-13
Save As, 1-8, 1-10, 7-2
Save Image As, 1-8, 1-10, 7-6
Save Picture As, 1-8, 1-10, 7-6

Search engine comparison, 3-10
Searchable directories, 6-3
Searching strategies, 5-11
Seattle Times, 8-11
Security, 4-13
Send Web pages, 4-12
Session Properties dialog box, 4-7
Shea, Virginia, 4-13
Smileys, 4-12
Society, 6-9
Sports, 5-10
Sports Illustrated home page, 1-10
Stock trading, 1-9
Style sheets 7-8, 7-15
Surfing, 6-2, 6-8

T

Teen smoking, 6-7
Thomas, 8-6
Title searching, 5-6
Training, 6-9
Trivia, 5-9
Tucows home page, 4-6

U

U.S. Census Bureau home page, 8-8

U.S. Treasury Department, 5-2
Universities, 5-10, 5-11, 6-2

V

Van Gogh, Vincent, 3-8
Virtual Jerusalem home page, 5-11
Virus, 8-9
Virus scan, 4-10

W

Watchdog groups, 8-9
Weather, 2-12
Web bugs, 8-11
Web Courier, 4-4
Web design, 7-13, 7-14, 7-15
Web tutorial, 1-6
White House home page, 1-8
White pages, 2-9
World Wide Web, History of, 3-11
World's fair, 3-9

Y

Yahoo home page, 2-5, 2-10, 2-11, 6-2, 6-7, 6-8, 8-4
Yellow Pages, 2-9

The Lab Activities for the World Wide Web, Annual Editions textbook will be updated each year. With your help next year's edition will be even more responsive to your needs.

Please make additional topic suggestions to the authors.

What labs in this Annual Edition need improvement, and how could they be improved to suit your needs?

What labs in this Annual Edition did you NOT cover that we may consider eliminating in the next edition?

At which educational institution did you use this book?

Thank you for your response! Please FAX this sheet to us at (650) 726-4693, or email it to us at scotjones2@aol.com or pruby@mail.astate.edu